The POWER of Assessing

Guiding Powerful Practices

Lisa M. Nyberg and Julie V. McGough

Arlington, Virginia

"Assessment is today's means of modifying tomorrow's instruction."

—Carol Ann Tomlinson

Dedicated to all learners who seek answers and love learning!

Claire Reinburg, Director
Rachel Ledbetter, Managing Editor
Deborah Siegel, Associate Editor
Andrea Silen, Associate Editor
Donna Yudkin, Book Acquisitions Manager

Art and Design
Will Thomas Jr., Director

Printing and Production
Catherine Lorrain, Director

National Science Teachers Association
David L. Evans, Executive Director

1840 Wilson Blvd., Arlington, VA 22201
www.nsta.org/store
For customer service inquiries, please call 800-277-5300.

21 20 19 18 4 3 2 1

Cataloging-in-Publication data for this book and the e-book are available from the Library of Congress.
ISBN: 978-1-68140-549-0
e-ISBN: 978-1-68140-550-6

Contents

Part 1: Why Is Assessing a Powerful Teaching Tool?

Part 2: How Do I Design Authentic Assessments to Meet the Needs of ALL Learners?

Contents

Part 3: How Does Metacognition Support Instructional Decision Making?

Building a Cognitive Environment

Part 4: Videos

Appendixes

Contents

Color Coding

Throughout *The Power of Assessing*, the text, illustrations, and graphics are color-coded to indicate the components of the instructional model.

Questioning is printed in **red**.

Investigations are printed in **blue**.

Assessments are printed in **purple**.

When thoughtful **questioning** is combined with engaging **investigations**, amazing **assessments** are produced—just as when **red** and **blue** are combined, **purple** is produced.

We've also provided links and QR codes to the NSTA Extras page where you can view videos related to content throughout the book. Visit *www.nsta.org/assessing* to access all supplementary content.

The Authors

Lisa M. Nyberg, PhD

Professor, California State University, Fresno, Fresno, California

Dr. Nyberg has taught for more than 30 years as an educator of students ranging from preschool to graduate school. Dr. Nyberg currently works as an elementary science and technology teacher educator at California State University, Fresno. Her teaching has been recognized through a number of awards and features, including the Presidential Award for Excellence in Science and Mathematics Teaching, the National Science Foundation Award for Excellence in the Teaching of Mathematics and Science, the Disney Channel Salutes the American Teacher, and the *PBS ScienceLine* and *Learning Science Through Inquiry* television series. Businesses have also acknowledged her work, awarding her the OMSI/Techtronix Award for Excellence in Teaching and the American Electronics Association Award for Science Teacher of the Year.

Dr. Nyberg's academic background includes a BA in biology, a BS in education, an MS in education, and a PhD in curriculum and instruction. She has served as a board member for NSTA's Preschool and Elementary Division and its Preservice Teacher Preparation Division. In addition, she has served on a number of advisory boards, including those of PBS and the Children's Television Workshop for the television show *Sesame Street*.

Dr. Nyberg has worked as a leader in professional development. She has written and edited books, articles, and grants in the areas of science education and communication, including an award-winning book, *How to Talk So Kids Can Learn* (Scribner 1995), which focuses on communication strategies that teachers and parents may use to optimize learning.

Julie V. McGough

Teacher, Clovis Unified School District, Clovis, California

Ms. McGough has taught for more than 20 years, working with students from preschool through fifth grade. She currently teaches at Valley Oak Elementary, in Fresno, California. She has also worked as an adjunct faculty member and a master teacher preparing new teacher candidates for California State University, Fresno. Her teaching has been recognized with the Crystal Award in her district, and she is a California state finalist for the Presidential Award for Excellence in Science and Mathematics Teaching.

Ms. McGough's academic background includes a BS in education and an MEd in gifted education. She is currently pursuing a PhD in science education. Ms. McGough works as a leader in professional development and serves on the curriculum design team for her school district. She served on the NSTA Preschool and Elementary committee and has published in the NSTA journal *Science and Children*.

The *Powerful Practices* Series

The Power of Assessing is the third book in the NSTA Press *Powerful Practices* series and focuses on modeling authentic assessment techniques. The series also includes *The Power of Investigating* (2017) and *The Power of Questioning* (2015), which was selected by educators and publishers of the Association of American Publishers as a REVERE award finalist.

PART 1
Why Is Assessing a Powerful Teaching Tool?

Connecting Questions, Investigations, and Assessments

Why Is Assessing a Powerful Teaching Tool?

Children engage in thinking and learning as they use questions to dig deeper, make connections, or investigate phenomena. Authentic assessments shine a light on learning for both students and teachers! How does a teacher build and maintain a learning environment that supports authentic assessment? How does a teacher plan, manage, and engage students in ongoing assessments? How does a teacher model reflective practices to help students develop metacognitive skills?

Assessments allow students and teachers to reflect on and celebrate successes along the learning journey. In *The Power of Assessing,* we invite you to explore the potential of authentic assessments as we illuminate students and teachers engaged in questioning, investigating, and assessing as a community of learners.

Light goes through some objects. How does light go through water?

My group is building a model animal. How can I use these materials to show the body of a hammerhead shark?

Why Is Assessment Challenging?

Assessment is essential every day. According to Tomlinson, "Assessment is today's means of modifying tomorrow's instructions" (2014, p. 17). Multiple types of assessments provide information to make thoughtful instructional decisions. District and state assessments may be required and offer a piece of the puzzle to guide student learning. However, daily classroom experiences offer essential insight into how a child thinks or processes information in context.

Authentic assessments provide purposeful opportunities for students to showcase evidence of their understanding in a variety of ways. Providing multiple access points through learning experiences in the classroom shines a light on student strengths and may provide signals to help the teacher guide instruction in a different way. Darling-Hammond (2010) writes, "The Student Assessment System should employ a variety of appropriate measures, instruments, and processes at the classroom, school, and district levels, as well as the state level. These include multiple forms of assessment and incorporate formative as well as summative measures … [and] consider and include all students as an integral part of the design process, anticipating their particular needs and encouraging all students to demonstrate what they know and can do" (p. 5).

Authentic assessments help students tune in to learning rather than check out. Engaging assessments support students, providing feedback to guide learning challenges. Authentic assessment helps students see themselves as successful, reducing the likelihood that they will act out. Over time students that have opportunities to show what they know through their strengths will continue learning rather than drop out.

All students deserve to have an opportunity to build skills and excel in classroom settings. Authentic assessments can help teachers guide students through positive academic interactions. Students will gain confidence and learn to persevere through challenging tasks when their unique learning styles are valued and a growth mindset is encouraged.

What Is Authentic Assessment?

Scan the QR code or visit: *http://static.nsta.org/extras/practices/assessing/video1.htm.*

What Is Authentic Assessment?

Authentic assessment includes students performing real-world demands by applying content in context (Wiggins 2006). Performance expectations may outline what students should be able to create, do, perform, explain, or build to communicate their thinking and learning. Assessments should be thought of as something that happens throughout the learning journey to scaffold experiences along the way (see Table 1.1 and Figure 1.1).

Table 1.1. Assessing at Multiple Points

How can I communicate what I want to say? Students may write something that does not make sense, or they may struggle to organize ideas. Formatively assessing student needs during the writing progression helps teachers scaffold experiences and guide learning.	
What does this look like up close? Students may think out loud or develop new questions to take learning to a deeper level. Assessing understanding as students make observations provides important information for the teacher when planning future lessons.	
What happens to light when it goes through water? Assessing students while conducting investigations helps students construct new questions and helps the teacher plan next steps to guide the learning of concepts. Understanding how light travels through water helps students understand why animals live in different depths of the ocean.	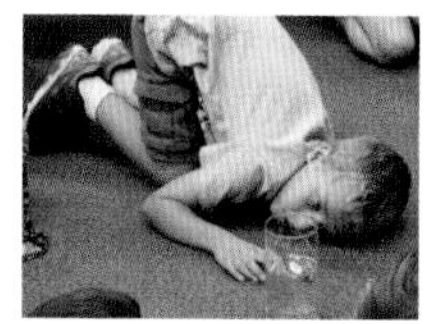
What happens to sound as it travels through water? As students ask questions about how whales communicate, they investigate how sound waves travel through water with a tuning fork.	
I wonder how I can find the answer? Students and teachers engage in metacognition to think about how they can find answers to their questions. Students learn to plan the next steps in their own learning.	
What sites might have pictures and videos? Knowing the right questions to ask will help students find important information more easily when using web resources.	
Where do I start? Assessment can be frustrating. It may not be a question of where to start, but rather a question of what to assess next or when to stop. Authentic assessment is ongoing and embedded in the learning process.	

Figure 1.1. Assessing at Multiple Points

Integrating Assessments

When assessments are integrated thoughtfully, a dynamic learning environment is created and is visible in the classroom (see Figure 1.3). The pictures below and on the next page show evidence that supported authentic assessments during a study of the ocean. Students engineered models to develop their understanding of the structure and function of ocean animals.

The coral reef display (Figure 1.2) shows student-engineered models of coral, anemones, a sea horse, a pufferfish, an eel, and a stingray.

The depths of the ocean display (Figure 1.4) shows student-created models of a dolphin, a hammerhead shark, jellyfish, a squid, and an anglerfish.

Figure 1.2. Coral Reef Model Animal Display

Figure 1.3. Visualization of the Create, Construct, Connect Concept

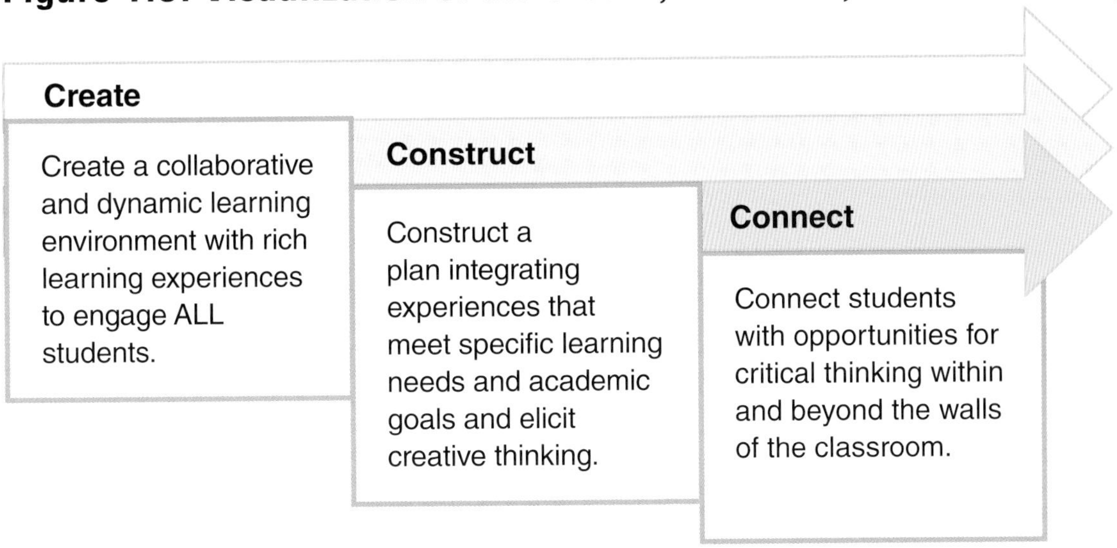

Create

Create a collaborative and dynamic learning environment with rich learning experiences to engage ALL students.

Construct

Construct a plan integrating experiences that meet specific learning needs and academic goals and elicit creative thinking.

Connect

Connect students with opportunities for critical thinking within and beyond the walls of the classroom.

Figure 1.4. Depths of the Ocean Model Animal Display

How Do Authentic Assessments Engage Students in Standards-Based Learning?

Authentic assessments illuminate student thinking and can be done through observation, questioning, conversation, and reflection on student work products or performance (NRC 2000). Pairing English language arts and mathematics standards with science standards creates opportunities for complex authentic assessments. The complexity of the task engages students at different levels, demanding application of knowledge and skills and synthesis of concepts. All students are unique and need occasions to show what they know in a variety of ways. Aligning assessments with academic standards, questioning strategies, and investigations creates a powerful practice to guide students and teachers in the learning process. (See Figures 1.5–1.7 on pp. 10, 12, and 14.)

Figure 1.5. Create, Construct, Connect

Create
Students collaborate to connect questions and content through reading, writing, speaking, and listening. Gathering information helps them prepare to create a three-dimensional ocean model while integrating literacy standards and science standards.

Construct
Students may use National Geographic videos to see anglerfish and eel in the ocean. Purposeful use of technology helps students construct new questions and plan next steps. Technology standards enhance integrated learning experiences. These experiences support complex authentic assessments (e.g., engineered models, reports, and presentations).

Figure 1.5. *(continued)*

Connect
Students think critically as they incorporate evidence from photographs and informational text to plan their three-dimensional model. Collaborative discussions engage students in questioning, investigating, and assessing as they think about how they will communicate the structure and function of the eel.

Engage in Three-Dimensional Learning

Powerful Practices integrate questioning, investigating, and assessing with paired standards to create three-dimensional learning experiences (McGough and Nyberg 2017). These experiences offer students and teachers the opportunity to engage in reflection, as ongoing assessment is embedded throughout the learning process. As students ask questions, teachers use questioning strategies to probe and extend learning. As students and teachers investigate answers to questions, formative assessment guides the next reading, discussion, or experience. Students and teachers may design performance assessments that shine a light on what they have learned. Students engage in reflection of their learning experiences, helping them to take ownership of their learning. Authentic assessment is meaningful to both students and teachers!

Figure 1.6. Powerful Practices: Primary

Powerful Practices: Primary

Second graders find and share evidence from text when comparing sea turtles to land turtles. Assessing student thinking during investigations helps teachers plan new experiences to scaffold learning. How are students interacting with text? How do students make sense of complex text when searching for information? Collaboration helps students put ideas together and plan new investigations to further their thinking. (See Table 1.2.)

Table 1.2. Examples of Possible Primary Standard Pairings

Grade	*Next Generation Science Standards* Disciplinary Core Ideas (NGSS Lead States 2013)	*Common Core State Standards* (NGAC and CCSSO 2010)
K	LS1.C: Organization for Matter and Energy Flow in Organisms All animals need food in order to live and grow. They obtain their food from plants or from other animals. Plants need water and light to live and grow.	ELA: W.K.7 Participate in shared research and writing projects. Math: K.MD.A.2 Directly compare two objects with a measurable attribute in common to see which object has "more of" or "less of" the attribute, and describe the difference. Math: MD.B.3 Classify objects into given categories; count the number of objects in each category.
1	PS4.A: Wave Properties Sound can make matter vibrate, and vibrating matter can make sound. LS1.A: Structure and Function All organisms have external parts. Different animals use their body parts in different ways to see, hear, grasp objects, protect themselves, move from place to place, and seek, find, and take in food, water and air. Plants also have different parts that help them survive and grow.	ELA: RI.1.1 Ask and answer questions about key details in a text. ELA: W.1.7 Participate in shared research and writing projects. Math: 1.MD.C.4 Organize, represent, and interpret data with up to three categories; ask and answer questions about the total number of data points.
2	LS4.D: Biodiversity and Humans There are many different kinds of living things in any area, and they exist in different places on land and in water. ETS1.B: Developing Possible Solutions Designs can be conveyed through sketches, drawings, or physical models. These representations are useful in communicating ideas for a problem's solutions to other people.	ELA: W.2.7 Participate in shared research and writing projects (e.g., read a number of books on a single topic to produce a report; record science observations.) ELA: W.2.8 Recall information from experiences or gather information from provided sources to answer a question. Math: 2.MD.D.10 Draw a picture graph and a bar graph (with single unit scale) to represent a data set with up to four categories.

Key: LS = life science, PS = physical science, ETS = engineering,
RI = reading informational text, SL = speaking and listening, W = writing

Figure 1.7. Powerful Practices: Intermediate

Powerful Practices: Intermediate

Third graders collaborate to build periscopes. Students reflect on their learning experiences about light and mirrors and make a plan to improve the design of their project. Students work together to solve problems while building the periscope. Reflecting on learning experiences helps students realize that assessment is an ongoing part of the learning process to plan and extend future learning. (See Table 1.3.)

Table 1.3. Examples of Possible Intermediate Standard Pairings

Grade	*Next Generation Science Standards Disciplinary* Core Ideas (NGSS Lead States 2013)	*Common Core State Standards* (NGAC and CCSSO 2010)
3	LS4.D: Biodiversity and Humans Populations live in a variety of habitats, and change in those habitats affects the organisms living there. ETS1.C Optimizing the Design Solution Different solutions need to be tested in order to determine which of them best solves the problem, given the criteria and constraints.	ELA: W.3.2 Write informative/explanatory texts to examine a topic and convey ideas and information clearly. ELA: RI.3.1 Ask and answer questions to demonstrate understanding of a text, referring explicitly to the text as the basis for the answers. ELA: RI.3.3 Describe the relationship between a series of historical events, scientific ideas or concepts, or steps in technical procedures in a text, using language that pertains to time, sequence, and cause/effect. ELA: SL.3.4 Report on a topic or recount an experience with appropriate facts and relevant, descriptive details. Math: 3.MD.B.3 Draw a scaled picture graph and a scaled bar graph to represent a data set with several categories. Solve problems using information presented in scaled bar graphs.
4	LS1.A: Structure and Function Plants and animals have both internal and external structures that serve various functions in growth, survival, behavior, and reproduction. PS4.A: Wave Properties Waves, which are regular patterns of motion can be made in water by disturbing the surface. When waves move across the surface of deep water, the water goes up and down in place; there is no net motion in the direction of the wave except when the water meets a beach.	ELA: W.4.1 Write opinion pieces on topics or texts, supporting a point of view with reasons and information. ELA: SL.4.5 Add audio recordings and visual displays to presentations to enhance the development of main ideas. Math: 4.G.A.3 Recognize a line of symmetry for a two-dimensional figure as a line across the figure such that the figure can be folded along the line into matching parts.
5	LS2.A: Interdependent Relationships in Ecosystems The food of almost any kind of animal can be traced back to plants. Organisms are related in food webs in which some animals eat plants for food and other animals eat the animals that eat the plants. Some organisms, such as fungi and bacteria, break down dead organisms and therefore operate as decomposers. Decomposition eventually restores some materials back to the soil. Organisms can survive only in environments in which their particular needs are met. A healthy ecosystem is one in which multiple species of different types are each able to meet their needs in a relatively stable web of life. Newly introduced species can damage the balance of an ecosystem.	ELA: RI.5.7 Draw on information from multiple print or digital sources, demonstrating the ability to locate an answer to a question quickly or to solve a problem efficiently. ELA: SL.5.5 Include multimedia components and visual displays in presentations when appropriate to enhance the development of main ideas.

Key: LS = life science, PS = physical science, ETS = engineering, RI = reading informational text, SL = speaking and listening, W = writing

How Does the Powerful Practices Instructional Model Work?

The Powerful Practices instructional model (Figure 1.8) engages students and teachers in the practices of inquiry and discourse through questioning, investigating, and assessing. *A Framework for K–12 Science Education* (*Framework;* NRC 2012) recommends that students develop the ability to call on the science and engineering practices to support their learning and demonstrate understanding. "Standards and performance expectations must be designed to gather evidence of students' ability to apply the practices and their understanding of the crosscutting concepts in the contexts of specific applications in multiple disciplinary areas" (NRC 2012, p. 218).

***Powerful Practices* Series:**

- *The Power of Questioning: Guiding Student Investigations* introduces the three-part instructional model and illustrates how to use types of questions to guide thinking and learning.
- *The Power of Investigating: Guiding Authentic Assessments* presents five types of investigations to help teachers and students transform two-dimensional learning to three-dimensional learning experiences.
- *The Power of Assessing: Guiding Powerful Practices* shines a light on learning by engaging students and teachers in authentic assessment for all students.

Different types of assessments put unique lenses on learning (Table 1.4, p. 25). When do I do assessment? Where do I do assessment? How and why should I do assessment? Assessing students through multiple access points provides opportunities for the teacher to be able to understand the view from multiple lenses, enabling all students to be successful in learning. Assessment choices must be multidimensional to meet the unique needs of learners.

Figure 1.8. The Powerful Practices Instructional Model in Action

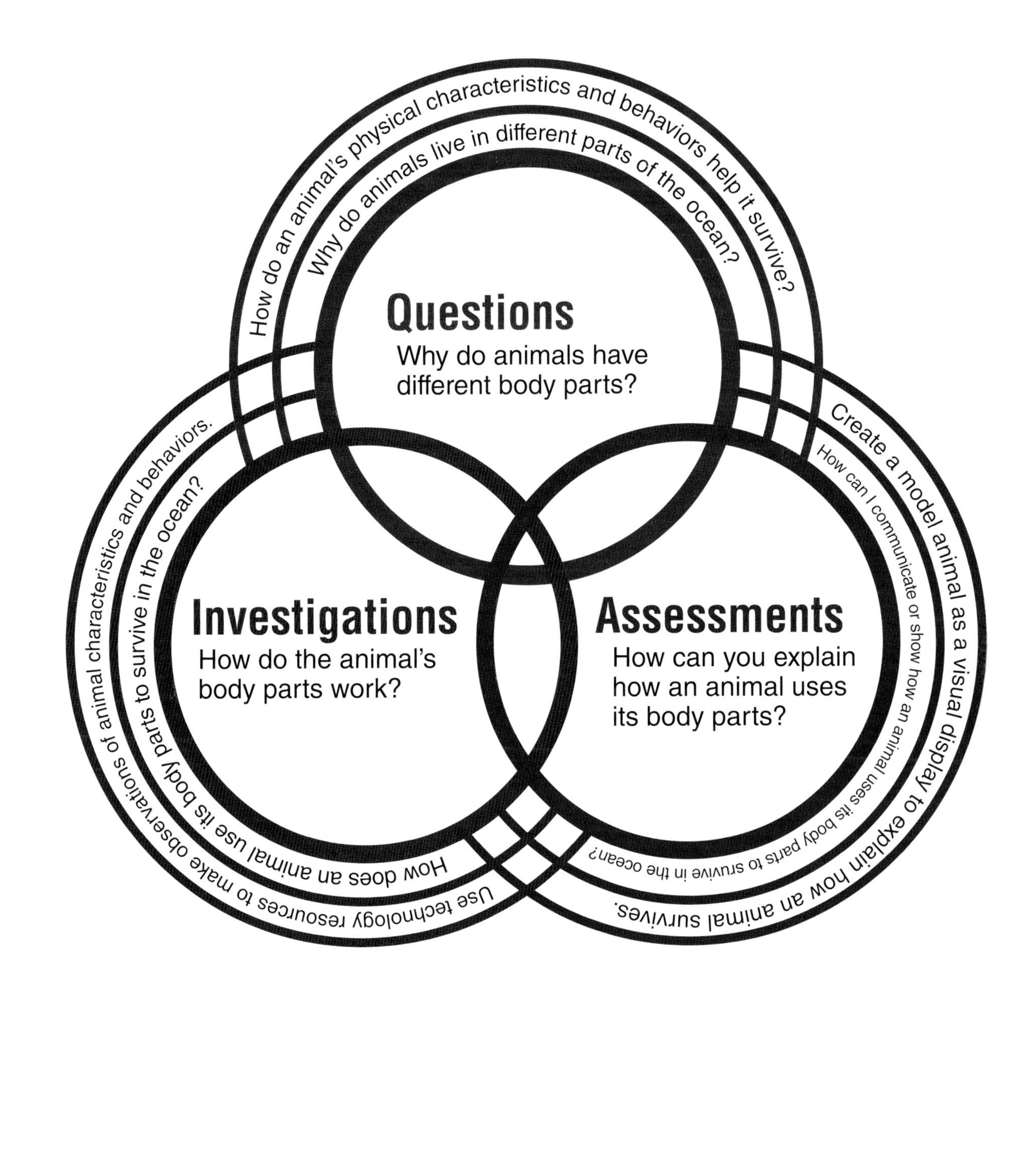

Integrate Questions, Investigations, and Assessments

Integrating questions, investigations, and assessments help students and teachers navigate learning through a model that is flexible and responds to the needs of the learners. (See Table 1.1, p. 6.) The model emphasizes three-dimensional learning experiences that spiral as new questions are developed, new investigations are planned, and ongoing formative assessment leads to new learning experiences. Because it is multidimensional, the Powerful Practices model supports cross-curricular literacy development as students construct understanding through the science and engineering practices, offering a dynamic interaction for learners and content.

In Figure 1.9, a group of three students form a research team to study the structure and function of a squid. The three-dimensional model builds on student strengths to meet student needs.

- In the red circle, the student used technology resources to observe the physical characteristics and behaviors of a squid.
- In the blue circle, the student used informational text to take a closer look at the tentacles of a squid.
- In the purple circle, the student recorded information from her group's findings.

The unit planning guide (Figure 1.10 [p. 20] and Figure 1.11 [p. 21]) provides a framework for building the integrated investigation lessons demonstrated in the text, pictures, and videos.

Figure 1.9. Integration of a Question, Investigation, and Assessment

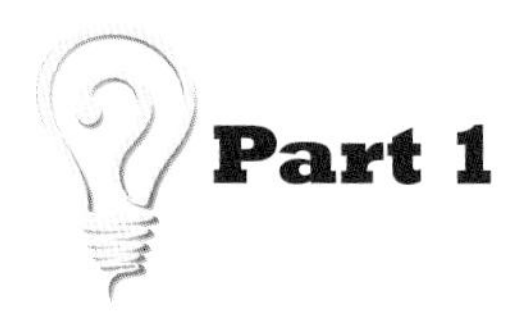

Using Unit Planning Guides

Figure 1.10. Using Unit Planning Guides

The unit planning guide in this example focuses on LS1.A Structure and Function as a life science standard integrated with PS4.A Wave Properties as physical science standards for light and sound. Weaving life science and physical science together throughout three-dimensional experiences creates multifaceted opportunities for students to shine a light on their thinking and learning. The unit planning guide can help you create a backward map to plan investigations that highlight complex concepts. Integrating curriculum supports individual learning needs and creates opportunities to make connections across disciplines. Each piece of learning contributes to student understanding and develops student confidence to engage in the process. Where do you want students to go? What do you want students to learn? How should you go about getting there? Listening to student questions and connections during the unit will help you to scaffold learning along the way. The unit planning guide helps the teacher consider cross-curricular opportunities and multiple intelligences to plan multifaceted experiences that tap into student strengths.

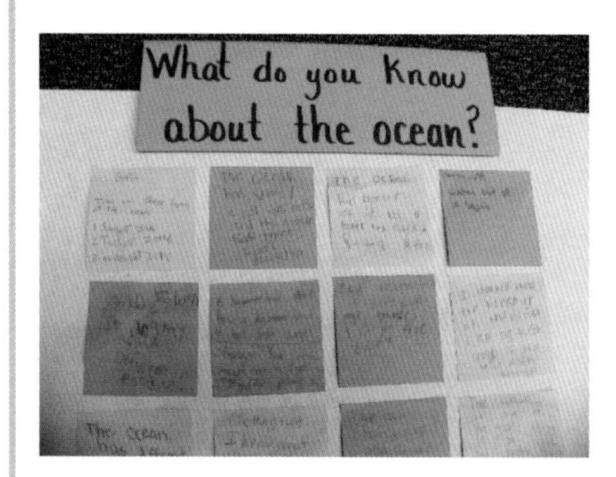

Preassess students by discussing a divergent question: What do you know about the ocean?

Formatively assess students during learning by reading journal entries, engaging in discussion, and asking clarifying questions.

The design, engineering, and presentation of a model ocean animal offers a summative assessment to interpret student understanding of the structure and function of ocean animals. Reflecting through group discussions and journal entries allows students to metacognitively describe their experiences.

Figure 1.11. Unit Planning Guide

Timeline:

Core Idea/Topic: Structure and Function

Concepts: External parts of animals

Questions to Drive the Inquiry

1. Why do plants and animals live in different parts of the ocean?
2. How do animals use light and sound to survive in the ocean?

Standards

NGSS

LS1.A.; All organisms have external parts. Different animals use their body parts in different ways to see, hear, grasp objects, protect themselves, move from place to place, and seek, find, and take in food, water and air; LS4.D Populations of organisms live in a variety of habitats and change in those habitats affects the organisms living there; PS4.A: Sound can make matter vibrate and vibrating matter can make sound.

***CCSS* ELA**

Key ideas and details (RI.1); **Craft and structure** (RI.4, RI.5); **Integration of knowledge and ideas** (RI.7); **Production and distribution of writing** (W.5, W.6); **Research to build and present knowledge** (W.7, W.8); **Comprehension and collaboration** (SL.1, SL2, SL.3); **Presentation of knowledge and ideas** (SL.4, SL.5, SL.6)

Student Questions

1. How does an animal use its body parts to survive in the ocean?
2. How does an animal's body (structure) create or receive sound to survive?
3. How does light go through some objects and not others?

Investigations

1. Research ocean animals through print and technology resources. Investigate how an animal uses its external parts.
2. Research ocean animals through print and technology resources. Investigate how animals use sound to communicate and sense predators or prey.
3. Observe sunlight outside the classroom. Explore light using flashlights and different types of objects inside the classroom.

Performance Assessment

1. Students will:
   a. Build a model of an ocean animal that illustrates how the animal's structure helps it survive in the ocean.
   b. Design a poster that explains how light or sound affects an animal in an ocean ecosystem.
2. Students will design a structure to illustrate and explain how light shines through some objects and not others.

Key: LS = life science; PS = physical science; RI = reading informational text; SL = speaking and listening; W = writing

Figure 1.11. (*continued*)

Cross-Curricular Connections

Science	Technology	Engineering	Mathematics	English Language Arts	Social Science	Art
• How do light and sound affect animals in an ocean ecosystem? • How does the structure of an animal help it survive? • How do plants and animals interact in an exosystem? • Animal dissection: squid; sea star; dogfish shark	• Digital microscope: View fish scales, lateral line, coral, sea stars, squid, etc. • Video resources: Aquarium web cams • Brochure/ pamphlet • Class blog: Report data and infrmation collected about plants and animals in an ocean ecosystem • Apps: Kids Discover, National Geographic, ARKive	• Build a periscope. • Design and build a model of an ocean animal. • Build straw pan pipes. • How do ships use sound waves?	• Graph the length of different types of whales. • How does the shape of an animal's body affect how it senses light and sound? • How does tooth shape show what kind of food an animal eats? • Create charts and graphs to organize the data collected from research and observations.	• Read informational text to explain key details about ocean animals. • Record observations and describe relevant details in journals. • Label diagrams and drawings. • Write reports communicating understanding of light, sound, and ocean animals.	• How are oceans around the world similar and different based on location? • Research and compare animals that live in different oceans. Locate geographic regions on a map. • How do plants and animals interact locally and globally? • How does a hurricane or other extreme weather occurrence affect an ecosystem?	• Fish printing: Create fish prints illustrating the external parts of a fish, scale patterns, and line. • Artist study (e.g., Robert Lynn Nelson): Paint ocean scenes illustrating the zones of the ocean. • Color value: Observe and create color value and shading by adding white or black to a base color. • Salt water painting

Multiple Intelligences

Linguistic	Logical-Mathematical	Visual-Spatial	Bodily-Kinesthetic	Musical	Interpersonal	Intrapersonal	Naturalistic
• Collaborative discussion • Reading • Writing • Brochures • Reports	• Graphs • Charts • Measurement • Organize data • Interpret data	• Videos • Technology applications • Hands-on investigations • Photographs • Drawings • Puzzles	• Hands-on tasks • Reader's theater • Plays • Outside explorations	• Songs • Chants • Poems	• Collective projects • Varied groupings	• Student choice • Reflection • Meaningful connections	• Animal care • Plant care • Work outside • Nature collections

Figure 1.11. (*continued*)

Resources

Coldiron, D. 2008. *Squid: Underwater World.* Minneapolis, MN: ABDO Publishing.
Coldiron, D. 2008. *Stingrays: Underwater World.* Minneapolis, MN:ABDO Publishing.
Coldiron, D. 2008. *Anglerfish: Underwater World.* Minneapolis, MN: ABDO Publishing.
Coldiron, D. 2008. *Eels: Underwater World.* Minneapolis, MN: ABDO Publishing.
Coldiron, D. 2009. *Sharks: Underwater World.* Minneapolis, MN: ABDO Publishing.
Coldiron, D. 2008. *Seahorses: Underwater World.* Minneapolis, MN: ABDO Publishing.
Coldiron, D. 2007. *Anglerfish: Underwater World.* Minneapolis, MN: ABDO Publishing.
Gibbons, G. 1995. *Sea Turtles.* New York: Holiday House.
Marsh, L. 2011. *National Geographic Kids: Sea Turtles.* Washington, DC: National Geographic.
Peterson, M. C. 2014. *Smithsonian Little Explorer: Coral Reefs.* Mankato, MN: Capstone Press.
Rake, J. S. 2009. *Beluga Whales Up Close.* Mankato, MN: Capstone Press.
Sexton, C. 2009. *Squids: Oceans Alive.* Minnetonka, MN: Bellwether Media.
Straz Center-National—Geographic Live! Brian Skerry Interview *www.youtube.com/watch?v=FJcLR86G7YQ*
The Traveling Turtle: Loggerhead Sea Turtle Critter Cam (video) *www.youtube.com/watch?v=8nmmYAtmag0*

Content Vocabulary

animal	external	lungs	senses
beak	feel	mammal	sensors
blowhole	fins	mollusk	skin
camouflage	fish	mouth	sound
carnivore	food chain	movement	sunlight
cephalopod	food web	ocean	system
claws	gills	omnivore	tentacles
community	habitat	parts	vibrations
crustacean	herbivore	predator	waves
ears	internal	prey	
ecosystem	life cycle	scales	
eyes	light	send	

Academic Vocabulary

change	display	informational text	record
clarify	dissect	journal	report
compare	evidence	label	resources
contras	explain	measure	similar
connect	facts	model	support
describe	fiction	nonfiction	technology
details	ideas	observation	thinking
different	illustrate	present	video
discuss	information	question	

Reflection

Integrating life science and physical science standards together increased the richness of the study of ocean animals. Students were able to dive deeper into the understanding of how animals use their senses to interpret the underwater world (e.g., hammerhead shark). Also, exploring light offered the opportunity to explore how and why different animals live at different depths of the ocean. Designing and building models of ocean animals helped students work together and understand complex ideas. Students were excited to see the classroom transform into an ocean environment and took pride in their work to show and teach others.

Developing Assessments: What Are Different Types of Assessments?

Assessments come in many forms and serve many purposes (see Table 1.4). Assessments help students and teachers connect concepts, think critically, and analyze what they know or don't know. Assessments also offer opportunities to plan new experiences to scaffold learning and progress to the next level. They can provide a means for students to think about their thinking along the way (metacognition).

Preassessment

Teachers often start with preassessment. To determine students' prior knowledge about a concept, the teacher may launch an investigation with a divergent question. As students share and connect ideas, the teacher may take mental notes, make an audio recording, or create a chart to record ideas of current student thinking and possible questions to investigate the concepts further. Preassessment also activates prior knowledge, student interests, and questions (see Figure 1.12, p. 26, and Appendix A, p. 94).

Formative Assessment

Formative assessment should be multifaceted and ongoing throughout the lesson or unit. The teacher can assess students during discussions, investigations, the writing process, close reading experiences, video presentations, and collaborative interactions among students while working. Making notes of the questions students are asking and how they are making connections may help to plan experiences that highlight complex information in a different way. Formative assessment helps students and teachers weave learning experiences together to construct understanding (see Figure 1.13, p. 27, and Appendix B, p. 96).

Summative Assessment

Multifaceted summative assessments make the learning accessible to all students. Diverse forms of assessment provide opportunities for all students to show what they know. Summative assessments should be planned to help students build on prior experiences and communicate a deeper level of understanding of the science standards and how ideas connect across disciplines (see Figure 1.14, p. 28, and Appendix C, p. 98).

Metacognition

Metacognition should be modeled by the teacher throughout learning experiences to help students engage in the process independently as they think about their own learning. Even young students can engage in metacognition. Valuing students' thinking helps students engage in the process and take ownership of their learning. For example, giving students opportunities to share their ideas of how to solve a problem highlights different perspectives and creates an opportunity for students to reflect on their thinking while connecting ideas (see Figure 1.15, p. 29, and Appendix D, p. 100).

Table 1.4. Types of Assessments

Assessment Type	Assessment Purpose	Assessment Questions	Example
Preassessment	Determine student's prior knowledge of the concepts Develop clarifying and probing questions and investigations Note misconceptions	What do you know about ____? How do you know? Where have you seen or learned about ____? How does this work? What would happen if _____? Can you explain your thinking?	Group discussion: What do we know about ocean animals? Discuss, draw, or write
Formative Assessment	Make and record observations Organize information Plan next steps	How does it change when ____? What happens when _____? What did you find?	Draw, label, or write what you know Group discussion Observe and ask questions
Summative Assessment	Organize and interpret information Synthesize information to create a model	What evidence from text can you find to support your thinking? How can your model show your understanding?	Write about the structure and function of your animal Create a diagram or label the parts of a photo
Metacognition	Ask questions about thinking and learning Reflect on the learning process and experiences	Why did you decide to use those materials? How does your model help you understand and explain the structure of the animal?	Group discussion Journal entry

Types of Assessments

Scan the QR code or visit: *http://static.nsta.org/extras/practices/assessing/video2.htm.*

Figure 1.12. Preassessment

What do students know?

Preassessment helps the teacher determine prior knowledge, misconceptions, and questions to investigate further.

The teacher and students discuss what they know about different ocean animals. What do students know about the animal's body structure and where it lives in the ocean (i.e., coral reef or different depths)? **How does the animal use its body parts to survive in the ocean?**

Criteria Checklist: What criteria will you use to assess student performance?

Teacher Action: Take notes during discussion.

- Students demonstrate in-depth, some, or no knowledge of the structure (external parts) of ocean animals.
- Students demonstrate in-depth, some, or no knowledge about where animals live in the ocean.
- Students demonstrate in-depth, some, or no knowledge about how animals use body structure to survive (function).

Figure 1.13. Formative Assessment

How are students progressing?

Formative assessment helps the teacher understand how the students are processing the content during the learning process.

Students may work individually or with a partner to draw and label an ocean animal. Clarifying questions may help the teacher gain understanding of the student's thinking. **How does the fin in your drawing help the animal move?**

mock trial
discussion
response app
four corners
journal
poster
post
graphic organizer
Formative Assessment
Socratic seminar
diorama
diagram
exit ticket
survey
think ink pair share
graph
artwork
timeline

Criteria Checklist: What criteria will you use to assess student performance?

Teacher Action: Review ocean animal drawings. Interview research teams.

- Students draw and label external parts (structure) of their ocean animal correctly.
- Students accurately describe where their research animal lives in the ocean.
- Students demonstrate through labeled drawings and discussion how their research animal uses its external parts to survive.

Figure 1.14. Summative Assessment

What did students know?

Authentic summative assessments offer students an opportunity to showcase their learning through evidence.

Students use technology to take a picture of their model and label the animal's external structure. They reflect on the model-building experience and articulate their understanding through speaking and writing. What materials did they use to build the model? **How does the animal's structure help it find food, communicate, or survive in the ocean?**

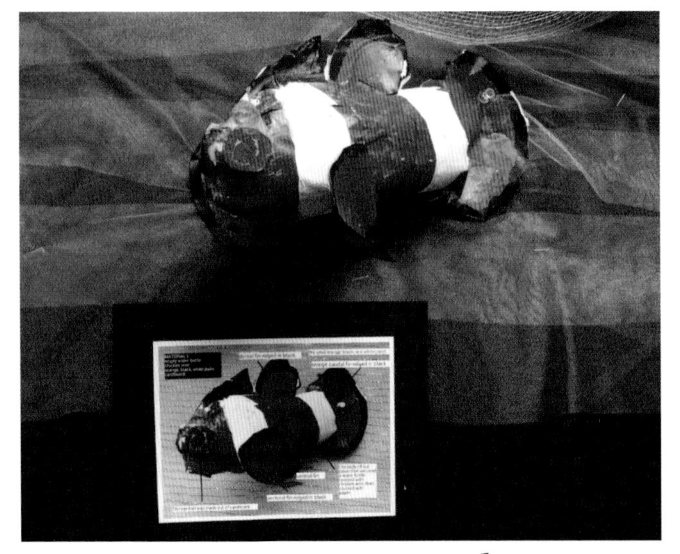

Criteria Checklist: What criteria will you use to assess student performance?

Teacher Action: Evaluate engineered models and digital models. Take notes during research team presentations to class.

- Students build an ocean animal model with evidence of all, some, or no external structures. Students correctly label the external parts on the digital model.
- Students clearly describe where their research animal lives in the ocean with evidence of all, some, or no correct terms.
- Students clearly, somewhat, or do not describe how their research animal uses its structure to help it survive in the ocean.

Figure 1.15. Metacognition

How will evidence help you make decisions?
Metacognition engages students and teachers in reflection of the learning.

Teachers model metacognition throughout learning experiences. Metacognition helps students make sense of complex information during the process of learning by thinking about connections, asking questions, and collaborating. **How does my model clownfish help me explain the structure?**

posts
journals
interviews
portfolios
discussions
progress reports
plus/delta
Metacognitive Assessment
response app annotations
goal setting

Criteria Checklist: What criteria will you use to assess student performance?

Teacher Action: Review reflections in student science notebooks. Listen to research teams reflect on the project.

- Students reflect on the engineering process of building a model of an ocean animal.
- Students reflect on their knowledge of key concepts about the structure and function of their ocean animal.
- Students reflect on the process of working collaboratively and presenting their work to an audience.

Preassessment Guide

A preassessment may start with a phenomenon, discrepant event, or discussion. Use divergent questions to engage students. For example, What do we know about ____? A whole-group or small-group discussion helps the teacher assess prior knowledge and prepares students to ask probing or clarifying questions during upcoming investigations. Preassessments help determine what current knowledge students bring to the lesson and lead to guided or independent investigations.

What **matcrials** do I need to engage students in this concept?

What **divergent question** focuses on the **criteria** for the task and helps me assess prior knowledge and misconceptions?

What questions will I use to keep the discussion going to help students **make connections**?

How will I use the preassessment information to **scaffold** learning experiences to meet the needs of ALL learners?

Preassessment Example

Topic: Structure and function of ocean animals

Preassessment: Small-group discussion

What **materials** do I need to engage students in this concept?

- Ocean animal photographs, ocean animal books, and National Geographic videos.

What **divergent question** focuses on the **criteria** for the task and helps me assess prior knowledge and misconceptions?

- What do you know about ocean animals?

What questions will I use to keep the discussion going to help students **make connections**?

- What are the parts of a ____ (specific ocean animal)?
- How does the animal use its body to move, hunt for food, and/or survive?

How will I use the preassessment information to **scaffold** learning experiences to meet the needs of ALL learners?

- Find further resources for small-group research. Gather recyclable materials for the next discussion to help students see, discuss, and design external structures of the ocean animal.

Formative Assessment Guide

Formative assessment is ongoing and multifaceted. Use clarifying, probing, and justifying questions to help students describe or explain their thinking. For example: How does the animal's body protect it from predators? Small group or independent investigations help students and teachers collaborate throughout the process of learning. The teacher can use this information to plan experiences that help students answer questions and ask new ones.

What **questions** will engage students in thinking about how to design their model according to the **criteria** for the task?

____

____

What **interactions** will help students make sense of the content?

____

____

How will I **guide** students to use resources and make connections?

____

____

How will I **scaffold** learning experiences for ALL learners and support further thinking?

____

____

Formative Assessment Example

Topic: Structure and function of ocean animals

Formative assessment: Small groups of three to four students will draw and label an ocean animal.

What **questions** will engage students in thinking about how to design their model according to the **criteria** for the task?

- What are the parts of your animal?
- How does the animal use its parts to survive in the ocean?

What **interactions** will help students make sense of the content?

- Students will use informational text, photographs, and National Geographic videos to observe the animal's body parts.

How will I **guide** students to use resources and make connections?

- Students will collaborate to work with and discuss information from text and video resources.
- Student groups will meet with the teacher to discuss questions and locate additional resources if needed.

How will I **scaffold** learning experiences for ALL learners and support further thinking?

- Gather materials and resources based on student questions and needs. Engage students in read aloud texts about the ocean habitat and animals. How are animals the same and different? How does their structure help them survive?

Summative Assessment Guide

Summative assessment may be multifaceted and performance oriented. Diverse assessments should be planned to help students build on prior experiences and communicate a deeper level of understanding of the science standards and how ideas connect across disciplines. For example: Design and build a model animal to communicate how the animal uses its structure to survive in its environment. Students may present the model and explain their knowledge and/or write about the experience.

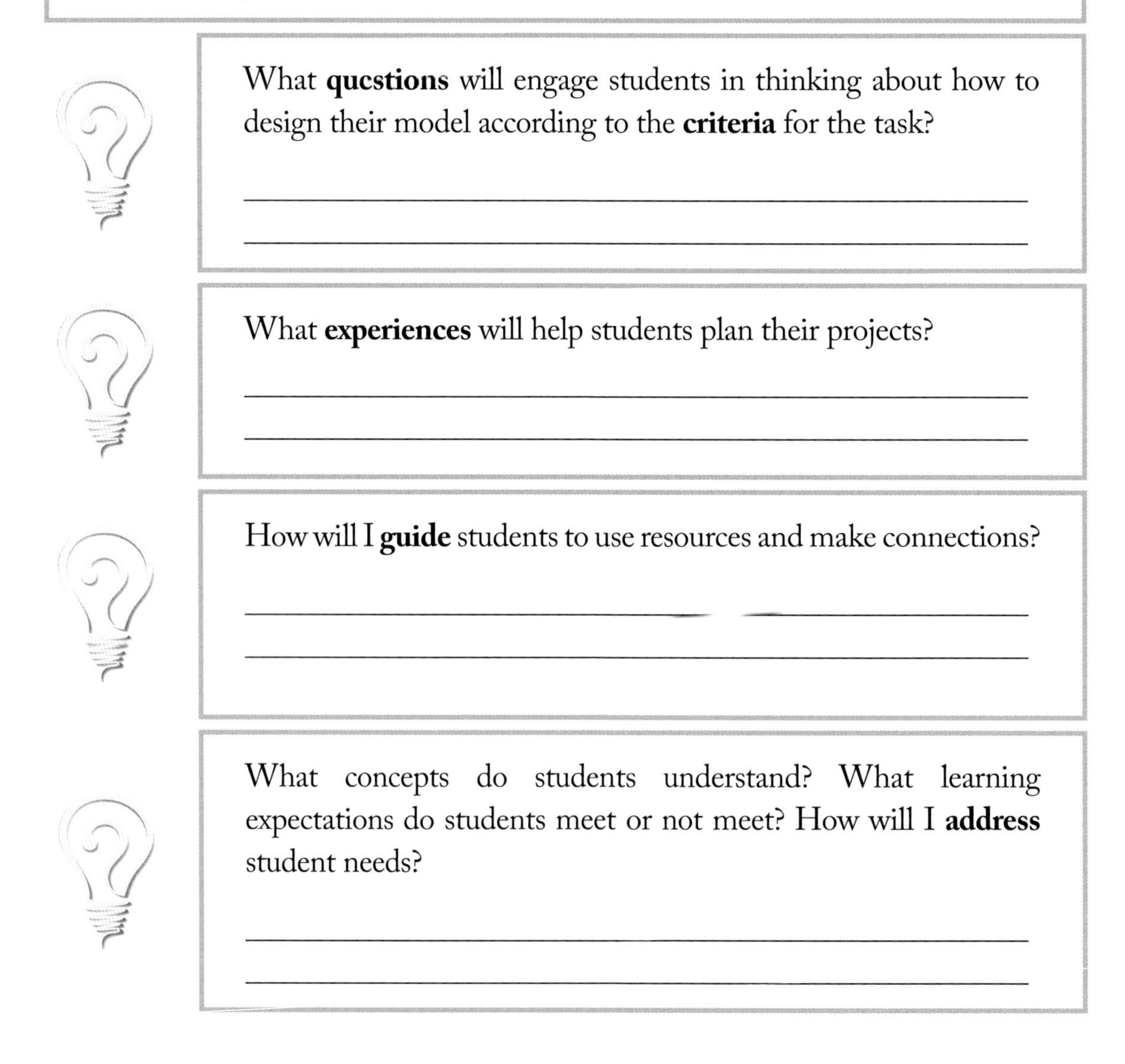

What **qucstions** will engage students in thinking about how to design their model according to the **criteria** for the task?

____

____

What **experiences** will help students plan their projects?

____

____

How will I **guide** students to use resources and make connections?

____

____

What concepts do students understand? What learning expectations do students meet or not meet? How will I **address** student needs?

____

____

Summative Assessment Example

Topic: Structure and function of ocean animals

Summative assessment: Design, build, and present a model of an ocean animal.

What **questions** will engage students in thinking about how to design their model according to the **criteria** for the task?

- How does the animal use its body parts to survive in the ocean?
- What materials can you use to show the parts?

What **experiences** will help students plan their project?

- Group discussion to prepare students to work as a team.
- Small group discussions to guide student planning.
- Discuss how different materials can be used.

How will I **guide** students to use resources and make connections?

- Students will use informational text, photographs, and National Geographic videos to observe the animal's body parts.

What concepts do students understand? What learning expectations do students meet or not meet? How will I **address** student needs?

- Students communicate understanding of animal structures through writing, models, and presentation of ideas.
- Determine new learning goals to address gaps or misunderstandings.

Metacognition Guide

Metacognition engages teachers and students in reflection of the learning process (see images on p. 37). These strategies may be used and modeled by teachers during learning experiences. For example, during class discussions teachers can model thinking out loud and help students learn to ask questions about their own thinking, which will help them to navigate their learning more independently. Teachers may reflect about thinking, planning, resources, and interactions. Students may reflect during discussions, during journal-writing experiences, and while creating work and/or projects.

What **questions** will engage students in thinking about their ideas and focus on the **criteria**?

- Can you explain your idea? What are you thinking?

What **experiences** will help students reflect on the concepts and process?

- Have small- or large-group discussions.
- Share ideas with others, then complement, connect, or question an idea you heard.

How will I **guide** students to use resources and make connections?

- Place students with similar interests together.
- Share books, videos, or other resources as connections are made.

How will students **reflect** on their experiences

- Reflect through small or large class discussions, writing, presentation opportunities, and/or while creating work or projects.

How can I put all of my ideas together in writing? How can I show my thinking with words?

I want to make my model like my picture, but how can I use the materials to show all the parts?

PART 2

How Do I Design Authentic Assessments to Meet the Needs of ALL Learners?

Engaging Students and Teachers in Assessments

How Do I Build an Assessment Plan Using Backward Mapping?

Working backward helps the teacher monitor progress as students build skills. Starting with a collaborative writing project that organizes the learning is a first step in preparing students to be able to write independently about any animal or topic. Group research and guided writing experiences scaffold learning to help students along the way. Building models helps students understand what they are reading and writing in a purposeful way. Students build confidence in independent writing as they use previously modeled strategies for a variety of topics. (See Figures 2.1–2.5, pp. 40–45.)

Figure 2.1. Backward Mapping

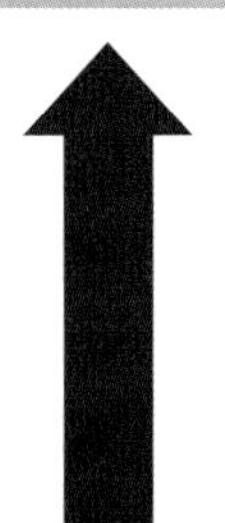

3 **Independent Writing:** Students apply research, reading, and writing skills to create writing on any animal or topic.

2 **Group Research:** Students work in groups to research the structure and function of an ocean animal and then design and build a model of the animal. Three-dimensional experiences include collaboration and discussion, video resources, reading, writing, building, and presenting.

1 **Collaborative Writing:** Students learn how to organize writing through a collaborative inquiry process that brings their ideas together to create a model writing product.

Building an Assessment Plan

Scan the QR code or visit: *http://static.nsta.org/extras/practices/assessing/video3.htm.*

Figure 2.2. Collaborative Writing

Collaborative Writing:

Students learn how to organize thoughts in writing through a collaborative inquiry process that brings their ideas together to create a model writing product. Integrating science and literacy standards engages students as they want to learn more and communicate their ideas. The collaborative writing lessons illustrated in the photo above and in Video 3 help students discover the purpose for organizing information when writing. Working collectively as a group guides students to evaluate information for accuracy and clarity. The teacher is able to formatively assess students throughout the process, making note of students who may need more scaffolding.

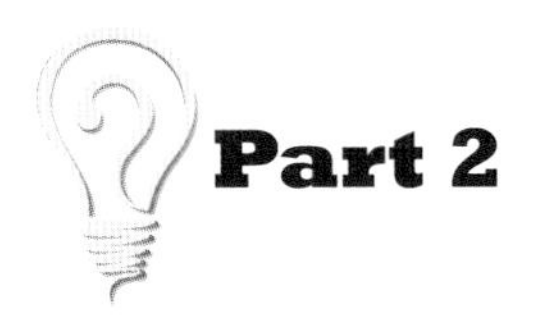

Figure 2.3. Group Research

Group Research:

Small groups engage in research using **National Geographic** (*kids.national geographic.com/animals*) and **Wildscreen ARKive** (*www.arkive.org*) to see photographs and videos of real ocean animals. Students are working on obtaining, evaluating, and communicating information. Visual resources provide context for student groups to discuss, collaborate, and plan their model animal. The teacher guides discussions through small-group focus investigations to help students make sense of the information. **Multiple experiences including observations, drawing, and designing a model will help support students as they write about their group's animal.**

Figure 2.4. Independent Writing

Independent Writing:

Rich and varied experiences provide a scaffolding of tasks to help students gain confidence in writing independently. Students apply research, reading, and writing skills to produce writing on any animal. Students are eager to research new topics as they obtain, evaluate, and communicate what they learn. Students refer to the collaborative writing as a model for their independent writing.

Backward Mapping Guide

Backward mapping helps you design learning experiences with the end in mind. What do you want students to be able to do? What experiences do you need to plan to meet the needs of all students and achieve writing goals?

First: What do you want students to learn, understand, or be able to do?

- What are the science and literacy standards?
- What is the science performance expectation connected to your content?
- What is the final product or expectation for learners?

Next: What evidence are you looking for that will help you plan experiences to scaffold learning to the next level? What resources will students use? (technology, text, tools)

- What will students do collaboratively versus independently?
- What will students create or build to help them understand the content?
- How do conversations, discussions, and questioning strategies help you identify evidence of learning?

Last: How will students apply their learning in a new context?

- What can students read, write, and learn next to extend their learning or provide independent practice?
- What experiences might you plan to meet the needs of students that did not reach the goal?
- How will you engage students that met the goal to continue learning?

Figure 2.5. Backward Map to Support All Learners

Meeting the needs of a diverse group of learners can be a challenge. The picture above shows a combination class of first and second graders including students with special needs such as autism, language learners, and other needs. What questions will you ask to assess what the students know? How will you use this information to plan and scaffold lessons that help all students reach identified learning goals?

How Do I Design Assessments for ALL Learners?

Students need opportunities to show what they know in a variety of ways and at multiple points along the way. Knowing what to look for and how to interpret what you see may guide you to plan new experiences. Questioning, investigating, and assessing help you guide learners and plan experiences that offer ALL learners the opportunity to construct understanding.

Universal Design for Learning guidelines suggest providing multiple means of engagement for the learner; providing multiple means of representation to present the content; and providing multiple means of action and expression so the students can show what they know (CAST 2011; Nelson 2014). (See Figure 2.6, p. 47.)

Gardner's Multiple Intelligences theory suggests that people have very different intellectual strengths. These strengths are very important to how people represent things in their minds and how they express their understanding (Edutopia 2009; Armstrong 2009). (See Table 2.1, p. 50.)

Consider the synergy of looking through the lens of multiple intelligences while providing multiple means of action and expression where students may show what they know in a variety of ways. Imaginative and innovative assessment design helps diverse learners and teachers access understanding. (See Figure 2.7, page 49.)

Assessments for ALL!

Scan the QR code or visit: *http://static.nsta.org/extras/practices/assessing/video4.htm.*

Figure 2.6. Universal Design for Learning

Engagement: The student is interested in how a real squid moves. He views an ARKive video to see the animal in an authentic context, and he reflects on how to add details to his group's model.

Representation: The student visualizes the squid's tentacles as he articulates how to use a bath mat to represent the suction cups. Small-group discussions offer opportunities to use vocabulary in context.

Action and expression: A group of three students uses multiple tools and types of media to communicate, design, and build a model squid.

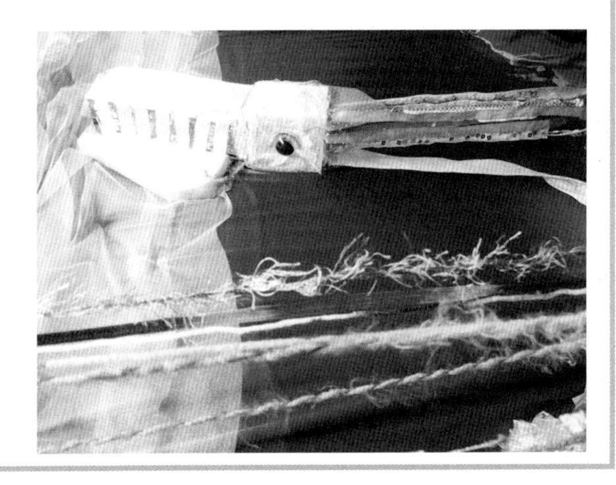

Figure 2.6. (*continued*)

Assessment Design:

Engagement through questioning: How does the structure of the squid help it to survive?

Representation through investigating: The student engages in discussions with the teacher and other students about the external parts of a squid. He reads and watches videos to identify shapes, features, and materials to design and build a model squid.

Action and expression through assessing: The student interacts with content, peers, and the teacher to think about and process information throughout the learning process. The student presents his learning and reflects on the experiences to communicate what he knows.

Figure 2.7. Assessment Design

Students learn how the structure and function of an animal help it find food, escape predators, and communicate in different parts of the ocean. Students may research ocean animals using nonfiction books and internet resources such as ARKive and National Geographic. After viewing photographs and videos, the student analyzes the structure of the squid by looking at shapes and recycled materials that could represent that design when building a model. Below left, the student's writing uses evidence from text to support his ideas; below right, squid model.

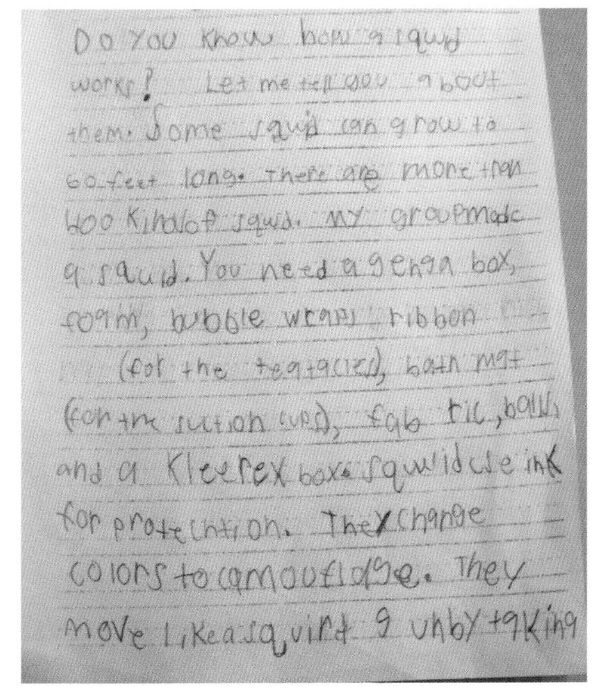

Do you know how a squid
works? Let me tell you about
them. Some squid can grow to
60 feet long. There are more than
400 Kind of squid. My group made
a squid. You need a gena box,
foam, bubble wrap, ribbon
(for the teatacles), bath mat
(for the suction cups), fab tic, balls
and a Kleenex box. Squid use ink
for protecntion. They change
colors to camouflage. They
move like a squid a uhby taking

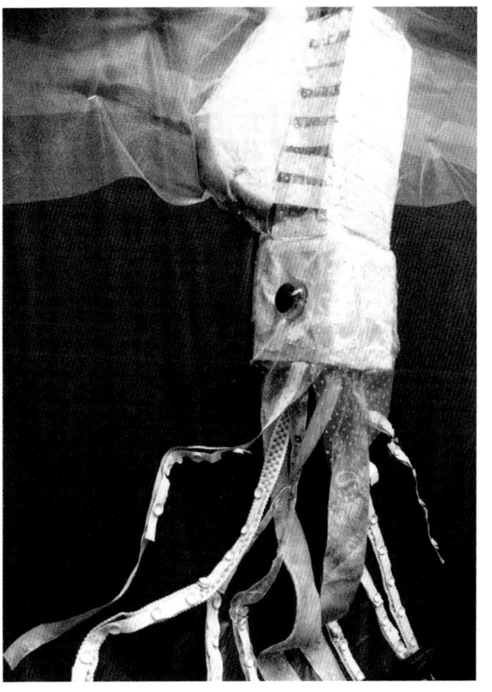

Table 2.1. Gardner's Multiple Intelligences

Multiple Intelligences	Questions	Investigations	Assessments
Visual-Spatial	How does light pass through objects?	Investigate different objects with a flashlight.	Draw the shadows of different objects.
Bodily-Kinesthetic	How do different drums make different sounds?	Investigate different sizes of containers (like cups, tubes, and tubs) to see how sound changes.	Build a sound toy that makes more than one sound.
Musical	What are the stages of the life cycle of a salmon?	Write lyrics to communicate the life cycle of a salmon.	Perform or create an audio or video recording of the song.
Interpersonal	How does an ocean animal use its parts (structure) to survive (function)?	Work with a partner or small group to investigate the ocean animal.	Build a model of an ocean animal.
Intrapersonal	How do sounds affect you?	Listen for different sounds around you. Make a list of the sounds you hear.	Record observations in a journal.
Linguistic	How do animals survive in different parts of the ocean?	Use videos and informational text to research an area of the ocean (e.g. coral reef or midnight zone).	Write a poem or make a brochure to illustrate an ocean animal's habitat.
Logical-Mathematical	How do different lengths of a tube change the sound? How can you put the sounds in order?	Cut tubes (or straws) to make different lengths. Investigate how the length affects the sound.	Demonstrate how pan pipes work by putting the straws in order.
Naturalistic	How do animals use sounds to communicate in the ocean?	Listen to audio recordings of beluga whales.	Write an explanation of how whales use echolocation.

Visual-Spatial Assessments

How can students see and touch?

Students learn through pictures and images.

Learn about jellyfish through nonfiction text and video resources. Create jelly fish using balloons, liquid starch, tissue paper, and yarn. This hands-on project is messy but provides a foundation for building three-dimensional models later in the ocean unit.

Other examples:

- Models, photos, graphics
- 2-D or 3-D art projects
- Slide show, mural, diagram

Visual-Spatial Assessment Criteria

How can you assess student understanding through visual-spatial learning experiences?

- Students increase understanding of building accurate 2-D and 3-D models through a scaffolding of experiences (e.g., jellyfish model art project provided a foundation for building a more detailed model to represent the external structure of an animal).
- Students communicate understanding of concepts learned from text and video resources about a topic as they create a visual representation (e.g., jellyfish art project above).
- Students engage in discussion using accurate terms and ask clarifying questions to further understanding while participating in the visual-spatial learning experience.
- Students design, draw, and build models to accurately represent concepts and terms.
- Students create an accurate slide show or other digital representation of concepts or terms.

Linguistic Assessments

How can students read, write, and communicate?

Students learn through words.

Most content areas are assessed linguistically in school. Providing students with rich visual resources like informational text and videos in addition to discussions leads to amazing writing. The writing about jellyfish in this photo communicates what the child has learned.

Other examples:

Journals report, brochure

Story, song, poem

Newscast, post, webpage

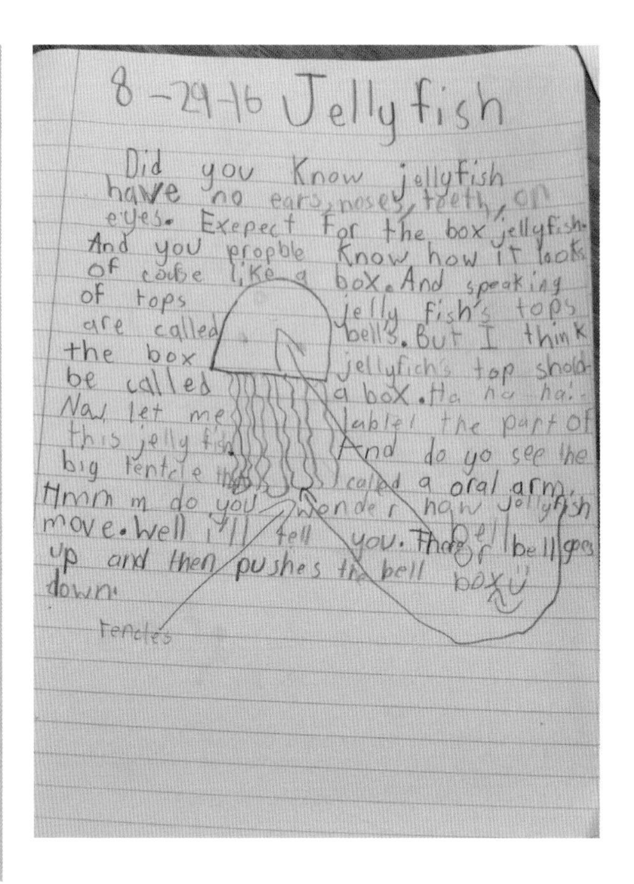

Linguistic Assessment Criteria

How can you assess speaking, listening, writing, and wordplay for evidence of understanding?

- Writing clearly communicates accurate facts about the topic (e.g., jellyfish writing above) using evidence from text, digital, or presentation resources.
- Writing communicates accurate information in an organized, logical order.
- Partner and group discussions convey understanding through the correct use of terminology, clarifying questions, and further thinking.
- Stories, songs, or poems communicate understanding of the topic through correct use of terms and further thinking.

Bodily-Kinesthetic Assessments

How can students move and manipulate?

Students learn through their whole body and with their hands.

Learning about different kinds of whales helps first and second graders apply math standards for measurement through a bodily-kinesthetic graph. First graders could visually see and order shortest to longest. Second graders were able to measure children and whales in feet.

Other examples:

- Dance, play, puppet show
- Clay, manipulatives, games
- Props, simulations, tableau

Bodily-Kinesthetic Assessment Criteria

How can you assess student understanding through physical movement activities?

- Students accurately represent concepts (e.g., student body graph above).
- Students accurately explain concepts (e.g., shortest to longest or measurement in feet) after participation in the physical activity.
- A dance, play, or puppet show communicates understanding through correct use of terms and representation of the concepts.
- Students communicate understanding through correct use of terms and representation of concepts while using manipulatives or playing games.
- Students engage in the use of clarifying questions to further understanding of concepts in relation to physical movement activities.

Logical-Mathematical Assessments

How can students calculate, organize, and solve problems?

Students learn through numbers and reasoning.

Investigating light engages third graders in geometry. Students created a periscope by constructing a cylinder and placing mirrors at 45-degree angles. Third graders collaborated to analyze how a periscope works.

Other examples:

- Graphs, charts, spreadsheets
- Patterns, analogies, timelines
- Puzzles, riddles, codes

Logical-Mathematical Assessment Criteria

How can you assess student understanding through logical-mathematical reasoning?

- Students correctly use tools to design models with accurate measurements (e.g., building a periscope with correct angles).
- Students accurately represent or interpret information in graphs or charts based on data.
- Students communicate understanding by accurately explaining concepts through patterns, analogies, codes, and so on.
- Students engage in the use of clarifying questions to further understanding in relation to puzzles, games, and other logical mathematical activities.

Musical Assessments

How can students sing, compose, or find rhythm?

Students learn through music and rhythm.

Writing a song about the life cycle of salmon engaged students linguistically and musically. Students communicated their learning through performance. They not only wrote the lyrics but also created a video, too!

Other examples:

- Rhyme, poetry, sounds
- Jingles, chants, songs
- Parodies, music videos

Musical Assessment Criteria

How can you assess student understanding through music, rhyme, and rhythm activities?

- Students communicate evidence of understanding through an accurate representation of concepts in a song (e.g., salmon life cycle song and video above).
- Students communicate evidence of understanding with accurate information in a poem, chant, or rhyme.
- Students engage in the use of clarifying questions to further their understanding in response to musical or rhythmic activities.

Naturalistic Assessments

How can students observe and interact with the natural environment?

Students learn through interactions with plants, animals, and nature.

Reflecting, writing, drawing, and recording observations in the school garden engages students with the naturalistic intelligence!

Other examples:

- Field notes, nature collections
- Animal care, gardening
- Environmental action

Naturalist Assessment Criteria

How can you assess student understanding through observation and interactions with the natural environment?

- Students write text, draw images, or create diagrams that clearly communicate accurate information about the topic (e.g., garden observation) using evidence from nature (e.g., measuring the physical growth of a plant) in addition to text and digital resources.
- Students communicate understanding through field notes or a nature collection to accurately represent changes over time (e.g., daily observation of animal metamorphosis) or explain observations (e.g., collection of leaves or rocks).
- Students communicate understanding of concepts by engaging in clarifying questions to further understanding and planning a project that positively impacts the environment (e.g., recycling project).

Intrapersonal Assessments

How can students reflect and connect?

Students conduct independent research. They reflect and make personal connections to the content.

Students enjoy having choice in their investigations. They may set goals to plan their course of study. This student chose to investigate how mirrors reflect light.

Other examples:

- Scrapbooks, collections
- Portfolios, memoirs
- Blogs, diaries, journals

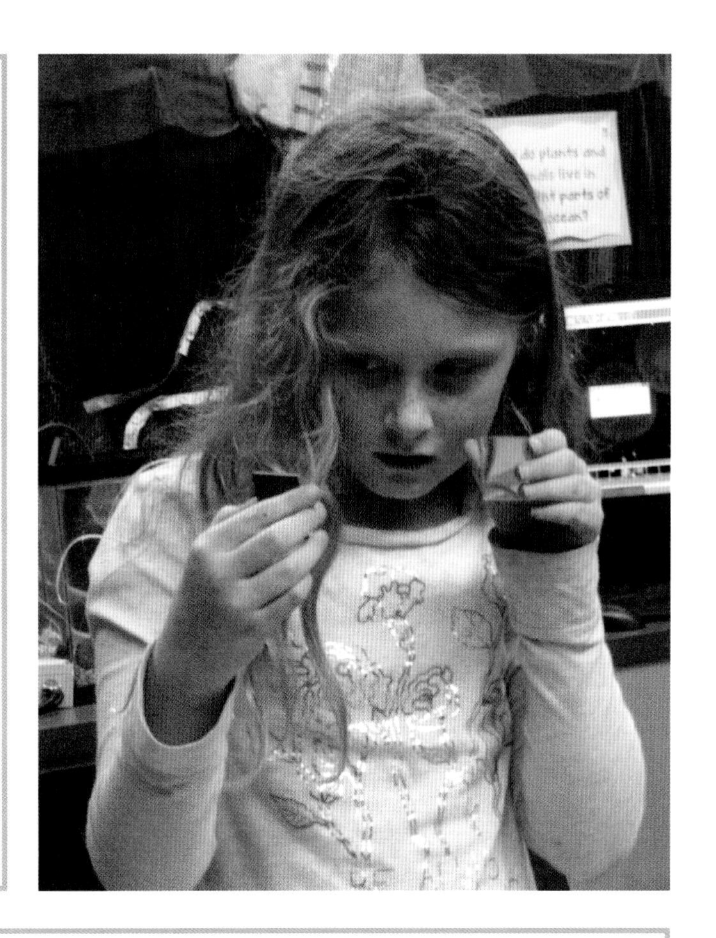

Intrapersonal Assessment Criteria:

How can you assess student understanding through reflections and connection?

Some students enjoy reflecting on experiences or working alone. This student chose to work alone while investigating how mirrors reflect light.

- Students create a blog, diary, or journal that accurately communicates an understanding of terms and concepts.
- Students' scrapbooks, collections, or portfolios communicate understanding by using accurate terms and organizing information in a clear, logical order.
- Students communicate through artwork or writing that clearly demonstrates an understanding of accurate information.

Interpersonal Assessments

How can students collaborate and connect?

Students learn through social interaction.

Many students learn well through collaboration with others. This project involved building a periscope with cylinder tubes, mirrors, and tape. Students helped each other solve problems and create working designs.

Other examples:

- Interview, debate, panel
- Team game, group work
- Survey, video, talk show

Interpersonal Assessment Criteria

How can you assess students through collaborative and social interaction?

- Students communicate understanding by using and explaining accurate terms during collaborative work (e.g., building a periscope required students to clearly articulate concepts to each other or their group).
- Students communicate understanding through social interactions (e.g., interviews or debates) by using accurate information and asking clarifying questions to further understanding.

Multiple Intelligences Guide

Multiple intelligences provide a way of thinking about assessment opportunities beyond chapter tests or tasks involving mostly paper and pencil. Think about the content students are learning through three-dimensional investigations. What is the best way for students to experience and communicate their understanding of the content? The photos on pages 51–58 illustrate examples of some of the many possibilities to assess student thinking during and after the process of learning.

How will I know which students benefit from **learning** through a different intelligence?

____

____

What tasks or projects will I **plan** to allow students to work in more than one intelligence area to address the **criteria**?

____

____

What **questions** will help me **assess** students according to the criteria during hands-on projects?

____

____

How will I **connect** tasks or projects from different intelligences within the same content?

____

____

Assessment Management Tips

How do I make sense of informal observations?

- Interact with individual students and small groups during work time. Ask questions to clarify student thinking, involving the learner in the process of explaining their ideas.
- Observe students as they interact with each other, making notes of questions and experiences that may help them to synthesize their ideas.

How do I record and organize questions and notes for planning?

- Organize a clipboard with student names to make notes of reading selections, interests, questions, etc.
- Keep a list of questions that may lead to further investigations.

How do I build on prior experiences?

- As students make connections, think out loud to celebrate the connection to prior experiences.
- Plan new investigations that help students make new connections.

How do I help students engage in reflection?

- Ask students why do you think that? How did you figure that out?
- Model reflection and metacognitive thinking throughout learning experiences.

Student journals are both a place for learners to process ideas and an ongoing record of progress. A variety of writing prompts will help teachers see how students are applying skills.

Independent reading time offers students an opportunity to develop and share new ideas. Student choice of reading materials helps teachers assess interest and guide new experiences.

Students may research concepts and organize ideas into digital documents or presentations. Storing work on Google Drive helps teachers assess student progress and plan new experiences to further thinking.

How Do I Reflect on Differentiation Through Student Learning Experiences?

Table 2.2 will help you use metacognition to reach all students.

Table 2.2. Metacognition at a Glance: Differentiation

What are students doing?	What are teachers thinking?
Speaking, listening, and asking questions	What prior knowledge do students have? What are student misconceptions? What are student interests? What are student learning styles?
Reading	How can I scaffold learning to help students access complex text? What experiences will help students find evidence in text that supports the criteria?
Writing	How can I scaffold learning to help students clearly communicate ideas in writing? What experiences will help students organize information in writing that supports the criteria?
Designing models	What resources do I need to provide to help students make sense of concepts? How can I scaffold learning to help students represent ideas clearly and support the criteria?

PART 3
How Does Metacognition Support Instructional Decision Making?

Building a Cognitive Environment

Metacognition engages students and teachers in reflection of the learning process—in essence, thinking about thinking. Metacognitive questions give teachers a better understanding of how students are processing content, the types of connections they are making, and how they are constructing knowledge. This insight provides a valuable tool when planning learning experiences to help children contend with big ideas. Metacognitive questions may include the following (see also Figure 3.1):

- How can you find the answer to your question?
- What steps do you need to take to help your group complete your project today?
- What other resources do you think you may need?
- How does your idea connect to the topic?

Modeling questioning and reflection throughout learning fuels active involvement in the assessment process, helping students see how to think out loud and internalize the language of questioning. Students need opportunities to learn from mistakes, persevere when learning is challenging, and think about content from different perspectives. Thinking out loud in the context of learning experiences helps students practice this valuable skill as they become independent learners.

Reflecting on learning experiences helps teachers communicate active questioning strategies and encourages students to use the same strategies as they reflect on their work. Ongoing authentic projects provide opportunities for students to reflect on learning and see progress over time. Making connections across the curriculum engages learners in various disciplines, applies knowledge and skills to complete authentic tasks, increases the complexity of thinking, and deepens understanding. Metacognitive questions support students and teachers as they wonder and learn together.

Figure 3.1. Student Metacognition

Student Metacognition

- How am I going to find out what I want to know? (researching, reading, writing, collaborating, building)
- What questions do I need to ask to figure this out? (clarifying, justifying)
- What can I do to help myself understand? (discussion, observation, work product)
- What learning goals am I trying to accomplish? (academic, social, performance expectation)
- How will I communicate or show what I am thinking? (drawing, writing, demonstration, presentation, discussion)

Metacognition: Evidence-Based Decision-Making

Scan the QR code or visit: *http://static.nsta.org/extras/practices/assessing/video5.htm.*

How Do I Use Metacognition to Make Instructional Decisions?

Reflecting on assessment opportunities throughout a learning sequence illustrates the connected nature of the learning process. Figures 3.2–3.12 and Tables 3.1 and 3.2 on the following pages provide a step- by-step annotation of both the collaborative turtle writing project and the model animal project. The teacher reflection boxes offer insight into the teacher's perspective on various learning experiences.

Figure 3.2. Teacher Metacognition

Teacher Metacognition

- What are students doing? (researching, reading, writing, collaborating, building)
- What questions are the students and I asking? (clarifying, justifying)
- How am I assessing? (discussion, observation, work product)
- How are the students self-assessing?
- What learning goals relate to the activity or investigation? (academic, social, performance expectation)
- What questions, investigations, or lessons are needed to scaffold learning to reach the learning goals?

How Do I Reflect on Assessment During a Collaborative Writing Project?

Table 3.1. Metacognition at a Glance: Collaborative Writing

What are students doing?	What questions am I asking?	How am I assessing?	What learning criteria relate to the activity or investigation?	What questions, investigations, or lessons are needed for scaffolding?
Investigating the topic	How do sea turtles grow and change?	Clarifying and justifying questions	Using resources, collaborating with others, gathering information, asking questions	Sea turtle resources, guiding questions
Reading and research	What facts can you find in the text about sea turtles?	Observation: How are students accessing text and text features?	Reading complex informational text Supporting ideas with evidence from text and resources	Close-reading lessons Shared reading to model finding evidence
Sharing and sorting ideas	Why do some ideas go together?	Observation: How are students organizing ideas?	Communicating clearly, organizing information logically, and asking relevant questions	Shared writing to model sorting and organizing ideas
Collaborating and writing	How did we put the ideas together?	Is the writing in a logical order? How do students reflect on the process?	Reflecting on work by: Clearly communicating facts in a logical order. Explaining the process with accurate details.	Small-group lessons to explain details and organize ideas

Figure 3.3. Launch the Topic

Launch the Topic: Begin the unit by sharing a short video of sea turtles entering the ocean after hatching. Ask students, what do you know about sea turtles?

- Preassess: What do students already know about sea turtles and other ocean animals?
- Preassess: What are student questions about sea turtles?

Teacher Reflection: A short video engages students and helps me share part of my summer vacation. The story of seeing Kemp's ridley sea turtles hatching one morning while I was walking on the beach sparks the beginning of a unit of study about the ocean and the many animals that live there. Sharing information about the organization that helps protect the nests provides a purpose for learning more. This helps me assess students' interest in the topic and their prior knowledge.

Figure 3.4. Reading and Research

Reading and Research: Gather books about sea turtles and distribute them to students. Give students time to read and look at the books. Ask students to write down two or three facts about sea turtles on a piece of paper.

Formative: Are students using text features to find information? Are students using pictures? How do students document evidence to support a fact or idea?

Teacher Reflection: This is an opportunity to walk around the room and interact with individual students and small groups to observe how they are interacting with informational text. Listening to students read, collaborate, and ask questions helps me think about additional resources that may be needed to answer student questions or plan investigations. Knowing that I am using the sea turtle inquiry as a way to model the process, I make notes of resources I may need when students begin to learn about other ocean animals in small groups.

Figure 3.5. Sharing and Sorting Ideas

Sharing and Sorting Ideas: Students bring the facts they wrote down to the group area to share. Ask someone to share a fact and place it on the floor. Continue allowing different students to share. What happens when facts are similar? How do students begin to make groups of facts?

Formative: Do students group facts in logical categories (e.g., life cycle, characteristics, diet, types)?

Preassess: Do students realize that facts within a group need to be in a logical order?

Teacher Reflection: This inquiry-based writing lesson helps students discover that facts and ideas can be grouped together. Engaging students in the process of writing provides scaffolding to help them become independent in other writing tasks. Collaborative writing provides a model for students to reference as they continue to write about other topics throughout the year. They will remember the experience because the level of engagement was high and the topic of sea turtles was meaningful.

Figure 3.6. Collaborative Writing

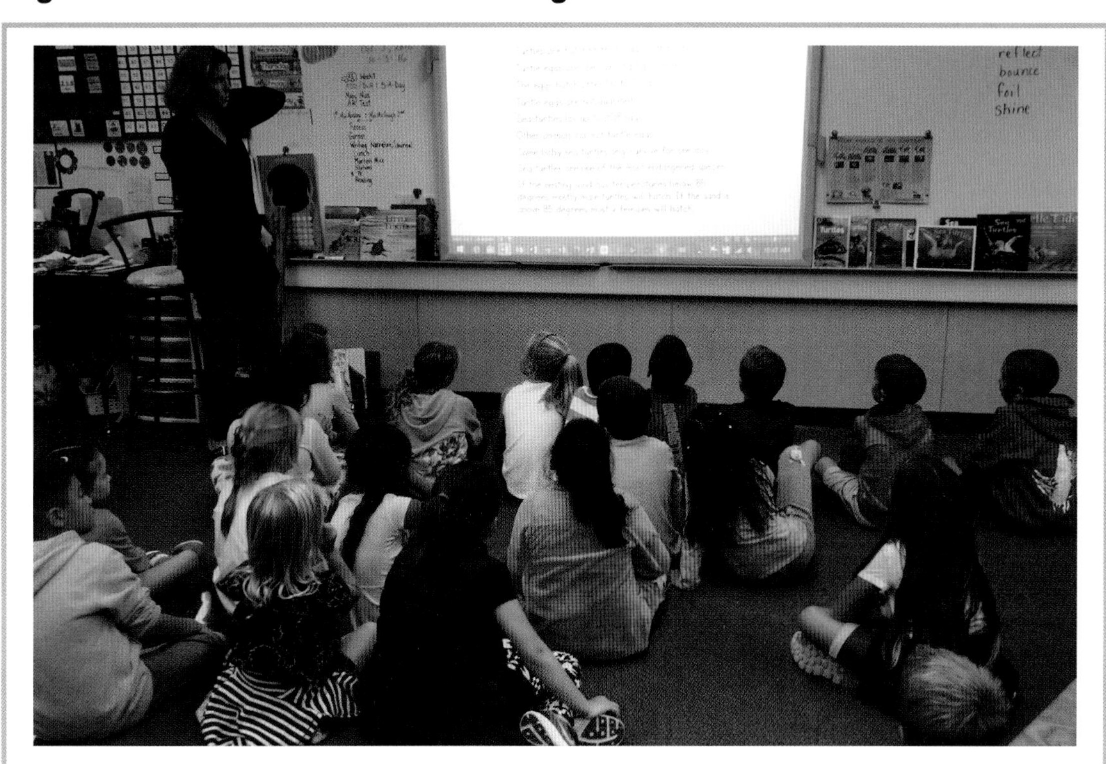

Collaborative Writing: Type or write the groups of facts that students organized (to make them easy to read and manipulate). Cut the facts apart and pass them out to members of the group. Have students read the facts out loud. Ask, are the facts in order? Should one fact come before another? Why or why not? Continue until the group agrees on a logical order for this section of text.

Formative: Do students want to add facts or details to clarify information?

Teacher Reflection: When students read the facts out loud they are actively involved in determining how the information should be organized in order for it to make sense. This process engages students in a purposeful reading and rereading of the text in order to revise and edit. After each section of text is organized, I type it to read through with the class and make final adjustments. Reflecting on the process helps students think through the steps and understand why each part is important. Metacognition is an important part of learning. This practice helps me see how students are thinking and shows that I value their opinions. Thinking out loud together helps all students learn along the way. Students learn to reflect on their own learning when working independently.

How Do I Reflect on Assessment as Students Engineer Models?

Table 3.2. Metacognition at a Glance: Engineering a Model

What are students doing?	What questions am I asking?	How am I assessing?	What learning criteria relate to the activity or investigation?	What questions, investigations, or lessons are needed for scaffolding?
Working as a team	How will we work together in small groups?	What are student's ideas about teamwork?	Communicate, collaborate, and solve problems effectively	Small-group games, discussions, and teamwork activities
Research and reading	What are the parts of the animal? How can we find out?	How are students accessing text and using resources to find information?	Organize information, gather ideas, and use evidence from text and resources	Guided research lessons; model finding evidence from resources
Designing and planning the model	What materials should we use?	How do students describe the structure of the animal?	Make a logical plan and list of resources relevant to the model	Guided small-group discussions and planning experiences
Building models	How will we build a model of this animal?	How do students solve problems while building the model?	Use key concepts and science vocabulary; ask relevant questions	Guided small-group work to design, build, and problem solve
Documenting and presenting	How will students communicate understanding?	How does the model help students articulate important information?	Clearly represent and communicate the structure and function of an ocean animal	Reflect and discuss as a small group. Use clarifying and justifying questions to help students explain concepts

Figure 3.7. Teamwork Discussion

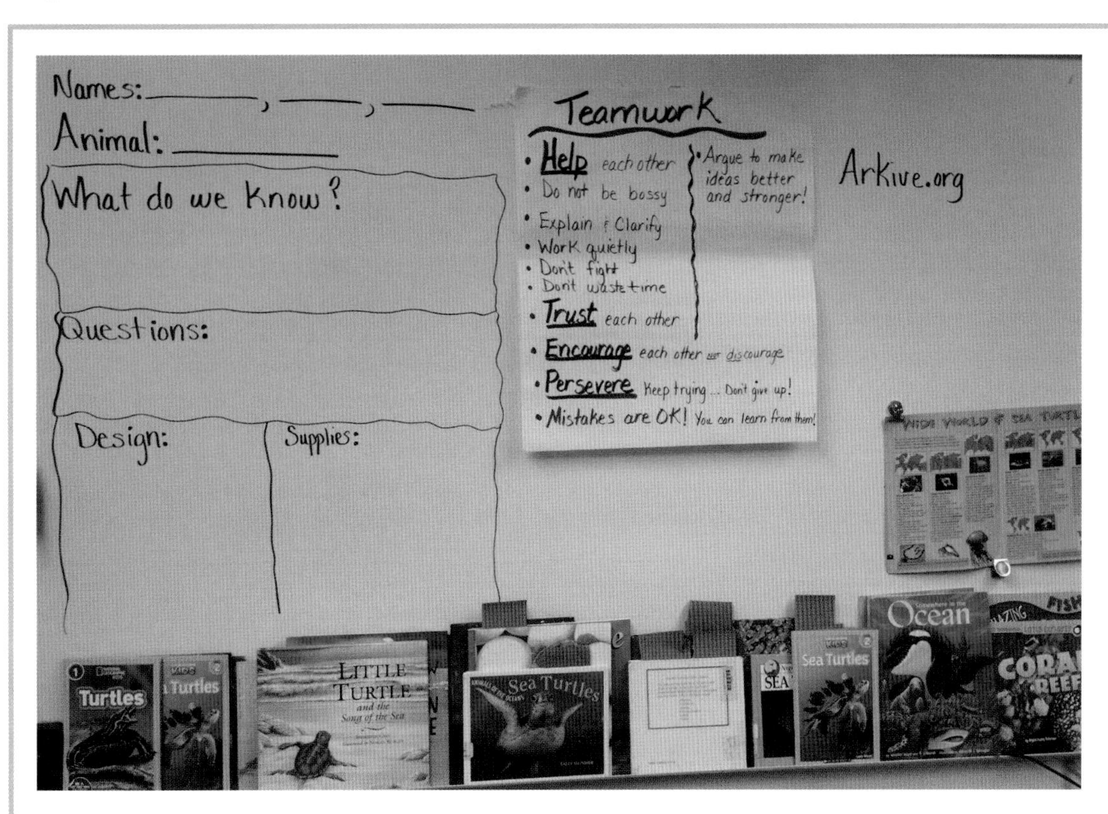

Teamwork Discussion: Begin the small-group work by setting the stage for successful teamwork. Ask students, "How can we work together as a team to accomplish goals? What do members of your team need to do?"

Preassess: What do students know about working together?

Teacher Reflection: Engaging students in a discussion about teamwork helps establish guidelines to accomplish goals. The teamwork discussion gave me great insight into how students think about taking turns and making mistakes. I asked if it was okay to argue with your teammates. Students were divided in their responses, but one student helped us understand how sometimes arguing can make your ideas stronger. This also led to a discussion about being open-minded and listening.

Figure 3.8. Research and Reading

Research and Reading: Gather books and digital resources about ocean animals and distribute them to small groups. Give students time to read and look at the resources. Ask students to record important information about the structure and function of their group's animal. They can also include diagrams and drawings to begin planning for the model animal.

Formative: Are students using text features to find information? Are students using pictures? How do students document evidence to support a fact or idea?

Teacher Reflection: Interacting with each small group helps me assess individual students and their needs as they work with complex text. I can provide guidance as needed or make a note to work with a student individually to reread text or reinforce skills.

Note: The formative questions are the same as during the collaborative sea turtle writing exercise. Students should be progressing in their ability to find and use informational text as they progress through the learning experiences.

Figure 3.9. Design and Plan the Model

Design and Plan the Model: Students bring the information they have recorded and any drawings of their animal. The teacher meets with each group to discuss what they have learned about the structure and function of their animal. Bring a container of recycled materials and objects to brainstorm possibilities of what to use when building the model animal.

Formative: How do students describe the structure of the animal? Are students missing important information that may be helpful when building their model? Are students making connections between the objects and the shape and structure of the animal?

Teacher Reflection: This small-group discussion is a great way to see how students are thinking about the structure and function of an ocean animal. Using objects and materials adds a tangible component as children describe, demonstrate, and explain how the animals move, eat, defend themselves, and hide. The objects also help students transition from a two-dimensional representation (photograph) to a three-dimensional representation (their model). Children show excitement in anticipation of the next step of building the model animal. The students and I make a list of any other materials that we may have at home that could be used.

Figure 3.10. Model Building

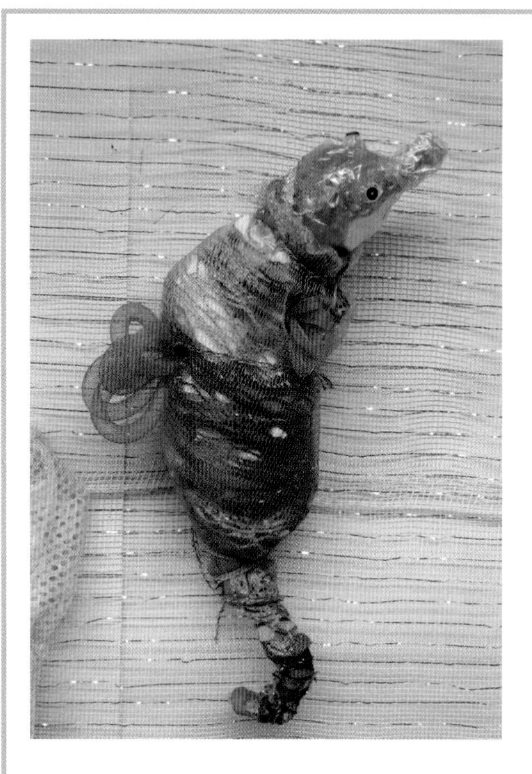

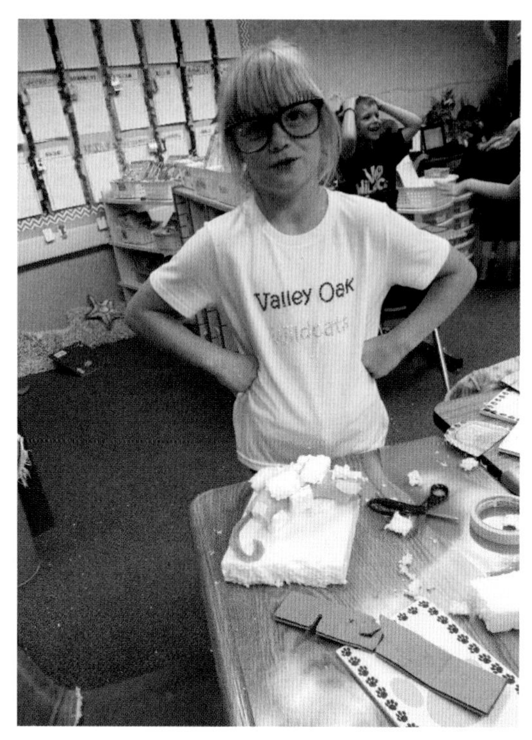

Model Building: Students work together with the materials and their notes to build a model of their ocean animal.

- **Formative:** How do the students organize their learning to construct a three-dimensional model?
- **Formative:** How are students showing understanding of the structure and function of the animal as they build the model?
- **Formative:** How do students solve problems while building the model?

Teacher Reflection: Interacting with small groups gives me the opportunity to assess individual student understanding of key concepts during the "mess" of learning. Students use science vocabulary as they collaborate with their team and build the model. This is a very dynamic and exciting part of the learning, offering students an opportunity to construct understanding in a meaningful way.

Figure 3.11. Digital Documentation

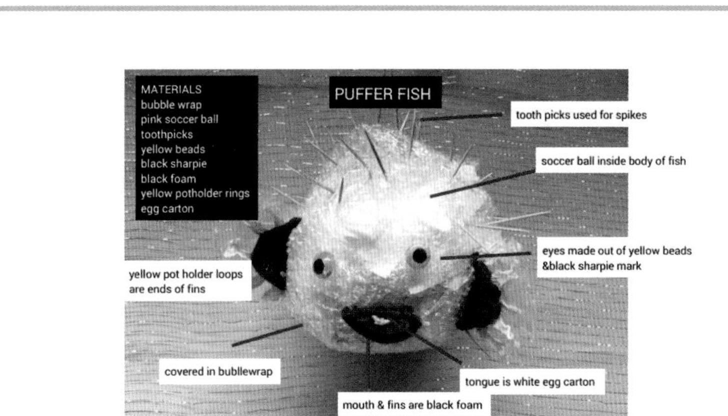

Digital Documentation: Each group will create a digital representation of their learning. The document will include a diagram of their finished model labeling the parts of the animal and explaining what they used for each part. It will also include photographs of the real animal and information about where the animal lives and how it survives in the ocean.

Metacognitive: How do students explain their thinking? How does building the model help students articulate important information?

Formative: How do students understand the structure and function of their ocean animal?

Teacher Reflection: Reflecting on the process helps students think through the steps and understand why each part is important. Metacognition is an important part of learning. This practice helps me see how students are thinking and shows value for the opinions of others. Thinking out loud together helps all students learn along the way. Students learn to reflect on their own learning when working independently.

This step also helps me see how students are processing important information as they initially reflect on the experience and complete their final product.

Figure 3.12. Presentation

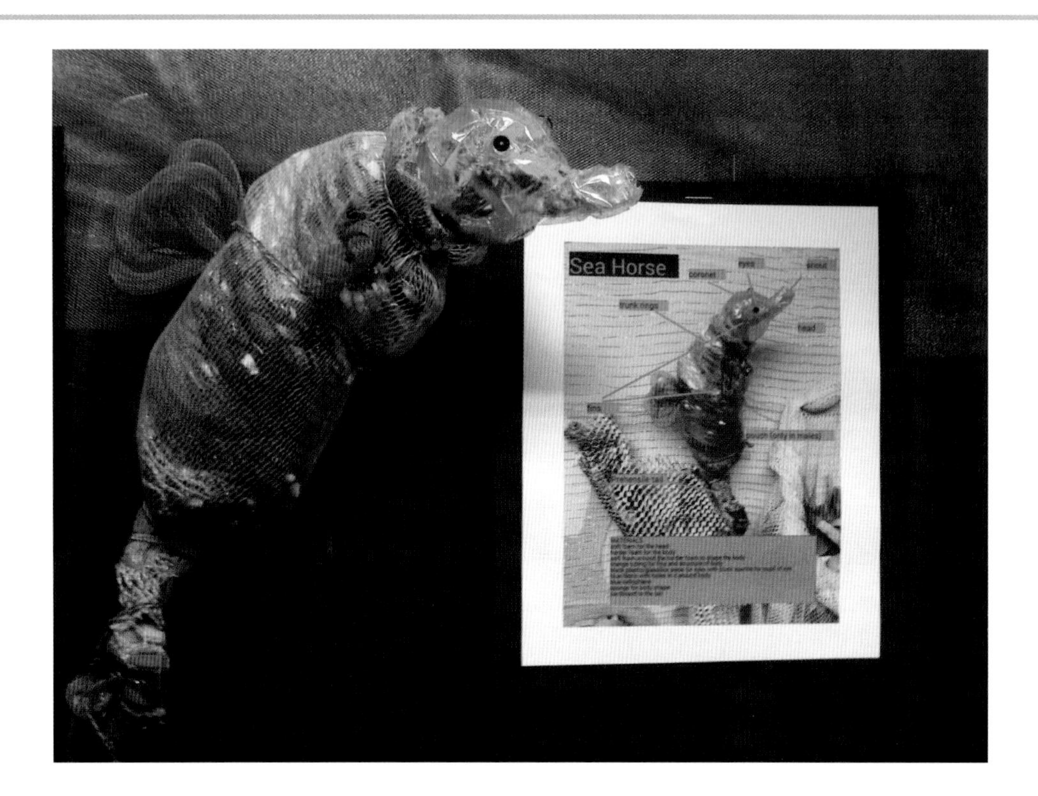

Presentation: Each group will present their model animal and the digital representation of their learning.

- **Summative:** Do students present their information completely? Are students able to answer questions from other groups, effectively articulating the information learned?

Teacher Reflection: The presentation phase of this project is a celebration of the entire process. Students articulate information they learned as they recount their experiences (research, planning, building, designing, etc.). Children are proud and excited to share this meaningful experience!

Reflect as a whole group on the process and experiences. How do students recount experiences? What were the struggles? What parts of the process did students find most rewarding?

The reflection as a whole group allows students to comment on group dynamics and offer suggestions for future learning experiences. This helps me plan to repeat aspects that went well and reevaluate aspects that were challenging. How can I improve on the experiences to offer better support or to deepen learning?

How Will I Design New Learning Adventures?

Assessments can be dynamic tools of learning. Assessments can be fun! Powerful questioning and investigating practices will lead you down a rich instructional path. When thinking of assessments some people refer to the quote, "I suppose it is tempting, if the only tool you have is a hammer, to treat everything as if it were a nail." Teachers have so many tools in their toolbox! Teachers are creative! Teachers are resourceful! Imaginative and innovative assessment design will energize your students and provide rich data to help you make instructional decisions. Your powerful instructional practices will change the lives of your students. How will you design new learning adventures to empower and inspire the children of today who will build our tomorrows?

PART 4
Videos

Video 1: What Is Authentic Assessment?

Focus: What does authentic assessment look and sound like?

Before the Video:

Traditional assessments require recall and recognition of a defined body of knowledge. Wiggins (2006) states:

> The point of assessment in education is to advance learning, not to merely audit absorption of facts … [With authentic assessment] students are tested on their ability to "do"the subject in context, to transfer their learning effectively.

Authentic assessment may take many forms. **With authentic assessments, students perform real-world demands by applying content in context.** Students learn to ask questions, conduct research, and show what they know with skills needed in life. Authentic assessments are developed to scaffold the needed skills and apply the content. Authentic assessments include tasks where students are required to construct and apply knowledge.

During the Video:

How does the teacher set up the room **to prepare** for authentic assessment opportunities?

What do you **see** in the classroom that provides **evidence** of student learning with authentic assessments?

What do you **hear** that provides evidence of student learning with authentic assessments?

After the Video:

How might you set up your room to **support** opportunities for authentic assessments?

What types of resources might you gather to **support** production of authentic assessments?

What would a visitor **see** in your classroom to provide evidence of student learning with authentic assessments?

What would a visitor **hear** in your classroom to provide evidence of student learning with authentic assessments?

What Is Authentic Assessment?

Scan the QR code or visit: *http://static.nsta.org/extras/practices/assessing/video1.htm.*

Video 2: Types of Assessments

Focus: What type of assessment will provide the evidence you need to make instructional decisions?

Before the Video:

This video features students studying science and engineering standards. In the first video sequence the first- and second-grade students have been learning about animals that live in the ocean. The students have been reading and writing about ocean animals. They have studied pictures and viewed media to see how their animal moves. The research teams have been learning how the physical characteristics and behaviors of the animal are linked to the animal's structure. They are exploring how materials may be used to build a model representation of their research animal.

In the second video sequence first graders are examining the concept of food chains. They are learning to read with informational text. They begin to connect the concepts of predator-prey relationships. They also are beginning to learn how food chains link to form food webs.

During the Video:

What **preassessments** do you see and how might they support lesson and unit planning?

What types of **formative assessments** do you see and how might they support instructional decision making for teachers and students?

What types of **summative assessments** do you see and how might they provide evidence to support evaluation?

How does **metacognition** support student and teacher decision-making?

After the Video:

Select a standard. What are the performance **goals** for the students?

What **preassessments** might guide your planning?

What **formative assessments** might guide teachers' planning and students' self-monitoring?

What **summative assessments** might provide an authentic evaluation?

How might you use **metacognition** to support your instructional decision making?

Types of Assessments

Scan the QR code or visit: *http://static.nsta.org/extras/practices/assessing/video2.htm.*

Video 3: Building an Assessment Plan

Focus: How can you build an assessment plan?

Before the Video:

Teachers may have a specific type of summative assessment or final project in mind. When teachers plan lessons with the end in mind (backward mapping) they carefully scaffold lessons to provide learning opportunities to build skills and provide the needed support for students to reach the goal.

This video demonstrates how to build an assessment plan. The teacher chooses to focus on the practice of obtaining, evaluating, and communicating the information. Her goal is to have the students be able to conduct independent research and produce writing on the ocean animal of their choice. She uses preassessments, formative assessments, and summative assessments to guide the learning opportunities for the students.

During the Video:

What evidence of the specific **goal** do you see?

How was **backward mapping** used to build an assessment plan?

How does the teacher use questions (**metacognition**) and student comments to build the assessment plan to guide the path of instruction?

How does the teacher sequence the lessons? How is the learning **scaffolded**?

What **preassessments, formative assessments,** and **summative assessments** do you see?

After the Video:

Select a specific **goal**.

What **preassessments** will help you know how to begin your instruction?

What **formative assessments** will help the teacher and students to assess skill development?

What **summative assessments** will provide an authentic student work product to demonstrate multiple facets of the content and skills?

How will the skill development be **measured**?

Building an Assessment Plan
Scan the QR code or visit:
http://static.nsta.org/extras/practices/assessing/video3.htm.

Video 4: Assessments for ALL!

Focus: How can you design assessments that are accessible to ALL of your students?

Before the Video:

The classroom is filled with diverse learners. Every child comes to the classroom with his or her own strengths and skills. Every child needs motivation to learn. Every child needs to be taught. Every child … needs to show what he or she knows. How do you design assessments that are accessible to all of your students?

Universal Design for Learning Guidelines suggest providing multiple means of engagement for the learner; providing multiple means of representation to present the content; and providing multiple means of action and expression so the students can show what they know (Nelson 2014).

Gardner's Multiple Intelligences theory suggests that people have very different intellectual strengths. These strengths are important to how people represent things in their minds and how they show what they've understood (Gardner 2013).

As you view the video, consider the synergy of looking through the lens of multiple intelligences while providing multiple means of action and expression where students may show what they know in a variety of ways.

During the Video:

How do students have the **opportunity** to "show what they know" in different ways?

What **evidence** of learning do you see?

What **evidence** of learning do you hear?

After the Video:

Select a standard. What is the student **performance expectation?**

How might you **design** an assessment to meet the performance expectation with action and expression with the lens of the following intelligences?

Linguistic:

Logical-Mathematical:

Spatial:

Bodily-Kinesthetic:

Musical:

Interpersonal:

Intrapersonal:

Naturalistic:

Assessments for ALL!

Scan the QR code or visit: *http://static.nsta.org/extras/practices/assessing/video4.htm.*

Video 5: Metacognition: Evidence-Based Decision Making

> **Focus:** How does metacognition guide the path of instruction and support independent student learning?

Before thc Video:

The focus of this video is **metacognition … or thinking about thinking.** Metacognition engages students and teachers in reflection. Metacognition is an ongoing process used to guide decision making and navigate instruction.

The first and second graders in this class are beginning small-group research investigations focused on an ocean animal followed by independent writing on an ocean animal of their choice. The teacher is setting the stage for learning by launching a discussion talking about what it means to work together as a team.

After the preselected small-group teams have drawn the name of their ocean animal, they begin to investigate the structure and function of the animal in preselected books and bookmarked websites. Smaller focus investigations follow where the teacher talks with the students about beginning to design their engineered model.

The National Geographic video clip featured:

You Tube: Straz Center-National Geographic Live! Brian Skerry Interview
www.youtube.com/watch?v=FJcLR86G7YQ

During the Video:

How does the teacher encourage **metacognition** on the topic of teamwork?

What **resources** does the teacher use to help the students transition from two-dimensional thinking (words and pictures) to three-dimensional thinking (models)?

How does the teacher **check for understanding** to guide her instructional decision-making?

After the Video:

Select a standard. What is the student **performance expectation?**

What opportunities for student **metacognition** might you provide?

How will you apply metacognition to check for understanding and guide your **instructional decision making?**

Metacognition: Evidence-Based Decision-Making

Scan the QR code or visit: *http://static.nsta.org/extras/practices/assessing/video5.htm.*

Appendixes

Innovative educators are always exploring new methods to add to their toolkits. When teachers use a rich variety of assessments, the learning environment pulses vibrant, strong, and healthy. The pages that follow (Appendixes A–D) provide expanded descriptions of the assessments introduced in the word cloud graphics on pages 26–29.

Explore preassessments and formative and summative assessment possibilities to provide a change of pace for your instructional day! Explore the use of metacognition for your students to support reflection and to help you navigate evidence-based decision-making opportunities. The assessment possibilities listed offer a springboard for planning. Your imagination will expand the possibilities as you create motivational assessment tools to inspire your students.

Appendix A: Preassessments

Preassessments help the teacher determine prior knowledge, misconceptions, and questions to investigate further. (See Figure 1.12, p. 26.)

Concept Map: The teacher asks the students to develop a graphic organizer to help students organize and represent knowledge of a concept. The concept bubbles and arrows show relationships between concepts.

Discussion: Open a discussion on a topic to access prior knowledge (i.e., what do students know about the topic and what misconceptions do they have?), determine interests, and explore questions of inquiry.

Fist to five: At the beginning of the lesson the teacher asks the students to self-assess their knowledge of the topic from fist (zero fingers: "I've never heard of the topic.") to five (all five fingers: "I could teach this topic").

Graffiti Wall: On a large piece of butcher paper or poster board, students use markers to write or draw whatever they know about a topic.

Hand Signals: Teach physical responses through hand signals. Students can signal agreement, disagreement, questions, offers of support, and so on.

Interview: The teacher may interview individuals or small groups to access prior knowledge, check for understanding, and determine interests.

KWL: On the board, chart paper, document cam, or computer:

The teacher may ask the students:

"What do you **K**NOW about ____________?"

"What do you **W**ANT to know about_______________?"

After studying the topic the teacher may ask:

"What did you **L**EARN about ________?"

Pretest: The teacher gives the students a written or performance test to assess the current level of skill and understanding. Quick Draw: The teacher asks the students to draw a quick sketch of a concept. For example: "With pictures, make a drawing of a food chain."

Quick Write: The teacher asks the students to write a quick explanation of a concept. For example: "What could you do at home to conserve water?"

Response App: Technology applications may be used to signal individual student responses. For example, programs like Kahoot!, Socrative, and Poll Everywhere.

Response Cards: The teacher may request that the students use cards to signal their responses (yes/no; true/false; agree/disagree). The whole class has an opportunity to respond individually.

Survey: Students may respond to a paper or electronic survey (for example, a Google Form).

Appendix B: Formative Assessments

Formative assessments help the teacher understand how the students are processing content during the learning process. (See Figure 1.13, p. 27.)

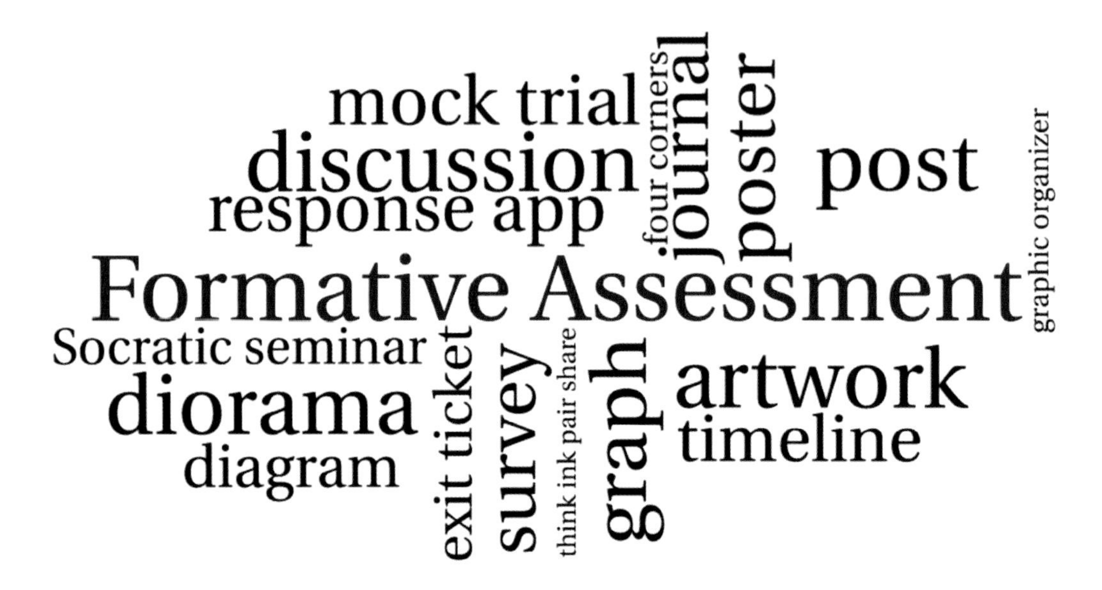

Artwork: Students create an art project (drawing, painting, sculpture) to illustrate a concept. For example: Paint a picture that represents the desert habitat with plants and animals. Discussion: Open a discussion on a topic to check for understanding.

Diagram: Students create a visual representation of a concept. For example: Draw a picture of the path blood travels in the body.

Diorama: Build a miniature three-dimensional model. For example: In a shoebox, create a model of a coral reef.

Exit Ticket: Students write a response to a question on a piece of paper and hand it to the teacher on the way out the door (on the way to recess, lunch, etc.). Some teachers have a taped-off grid with the student's ID on the wall or window, and the students must place a sticky note on the grid as thier exit ticket to leave.

Four Corners: Students are directed to move to the corner of the room that represents their response to a multiple-choice question.

Graph: Create a graphic representation to illustrate concepts. For example: Plot a graph illustrating the rainfall in an area over one year.

Graphic Organizer: Use a graphic organizer like a concept map. The teacher asks the students to develop a graphic organizer to help students organize and represent knowledge of a concept. The concept bubbles and arrows show relationships between concepts. Journal or Notebook: Students write a response or reflection in their journal or notebook. The response may be written in a paper notebook or it may be an electronic journal entry.

Mock Trial: Simulate a trial. For example: There is an oil spill in the ocean. Conduct a trial examining the issues between an animal rescue organization and the oil company.

Post: Students post a response to a prompt. The response may be in the form of writing on a wipe-off board, an app, or on a social media site.

Poster: Create a graphic representation of a concept. For example: Design a poster to persuade people to conserve energy.

Response App: Ask questions (like a pop quiz) to signal individual student responses. Use programs like Kahoot!, Socrative, and Poll Everywhere.

Socratic Seminar: The teacher asks a series of open-ended questions to stimulate a creative dialogue where students develop critical thinking skills as they listen to each other's comments and contribute to a collaborative discussion.

Survey: Students respond to a survey to check for understanding or position on a topic.

Think Ink Pair Share: Students first think about a response; then write down their ideas; then share their ideas with a partner. It is an assessment that encourages the student to develop and refine their ideas before sharing their thoughts.

Timeline: Students develop a timeline to represent a concept. For example, chart significant life events of a famous scientist.

Appendix C: Summative Assessments

Summative assessments help the teacher understand how students use evidence to showcase their learning. Summative assessments are major projects that highlight many different standards. (See Figure 1.14, p. 28.)

Collection: Gather materials that represent a concept in a collection. For example: Gather a collection of different types of rocks. Sort them into categories.

Debate: Students argue a point of view on a given topic. For example, "Should the rain forest be cut down?" Positions: Animals of the rain forest /Villagers who live in the rain forest /Timber workers/Environmentalists.

Engineer Model: Build a visual representation of a concept. For example: Design a three-dimensional model of an animal found in a coral reef. Design a two-dimensional diagram that illustrates the external structures and respective functions of those parts.

Environmental Action: Students perform a service-learning project where they plan and implement a service to the environment. For example: recycling, playground or school grounds renovation, public service announcements, etc. Research of Choice: Students have an opportunity to choose a research topic of choice, research and write on the topic, and then present the information.

Examination: Students complete a written or performance test on a topic to demonstrate what they have learned.

Multimedia Project: Students develop a project using a variety of media possibilities. For example: PowerPoint; Keynote; Prezi; iMovie; NearPod, etc.

Mural: Students paint or draw a collaborative art project to illustrate a concept. For example: Paint a mural showing ways students can protect the environment.

Music Video: Students research, write, perform, and produce a music video that communicates what they have learned on a topic.

Newsletter: Students develop a newsletter to share what they have learned with a larger audience (i.e., students, parents, school community).

Play: Students research, write, and perform a play to communicate what they have learned to a larger audience.

Puppet Show: Students research, write, and perform a puppet show to communicate what they have learned to a larger audience. Report or Presentation: Students share research and findings to an audience.

Scrapbook: Create a scrapbook representing a concept. For example: Design a habitat journal or scrapbook that illustrates observations of a specific neighborhood habitat.

Song: Students write the lyrics to a rap, chant, or song to teach a concept. For example: Students can rewrite the lyrics to a popular song to describe potential or kinetic energy.

TV or Radio Broadcast: Students develop and present a TV or radio broadcast. For example: Students record a news broadcast like *Good Morning America* for a larger audience.

Webpage: Design a webpage to share your knowledge with a larger audience.

Appendix D: Metacognitive Assessments

Metacognition engages teachers and students in reflection of the learning. Metacognitive assessments help teachers make instructional decisions. (See Figure 1.15, p. 29).

posts
journals
interviews
portfolios
discussions
progress reports
plus/delta
Metacognitive Assessment
response app
annotations
goal setting

Annotations: Students provide/receive comments on their work. They may make notes on their own work to self-assess and improve. They may provide peer feedback. They may respond to a teacher's feedback.

Discussions: The teacher and students have a discussion to analyze a product or process. For example: The teacher talks with the students about the characteristics of a strong work team. Goal Setting: Students set specific performance goals to target content and processes.

Interviews: Two people have a discussion to reflect on the product/process and make a plan for improvement. For example: Two students may serve as peer editors to review each other's work and give constructive feedback.

Journal or Notebook Reflections: Students write or draw in their journals or notebooks. What content and processes have they learned? What questions do they have? How could they improve their performance in the future?

Plus/Delta: This is a method of reflection where the participants indicate the "plus" (What is working.) and the "delta" (What action item are needed?).

Portfolio: Students select assessments that represent their work. This body of work may support student/parent/teacher conferences and opportunities of metacognition.

Posts: Students/teacher post reflective comments on progress on a product or process. For example: Students respond to comments from a teacher or peer.

Progress Reports: Students and/or the teacher analyze checkpoints to chart progress on a final project.

Response App: Students reflect on a concept, product, or process via a response app.

Support Resources

Children's Books

Cane, S. 2000. *Little turtle and song of the sea*. Brooklyn, NY: Interlink Publishing Group.

Coldiron, D. 2008. *Anglerfish: Underwater world.* Minneapolis, MN: ABDO Publishing.

Coldiron, D. 2008. *Eels: Underwater world.* Minneapolis, MN: ABDO Publishing.

Coldiron, D. 2008. *Seahorses: Underwater world.* Minneapolis, MN: ABDO Publishing.

Coldiron, D. 2009. *Sharks: Underwater world.* Minneapolis, MN: ABDO Publishing.

Coldiron, D. 2008. *Squid: Underwater world.* Minneapolis, MN: ABDO Publishing.

Coldiron, D. 2008. *Stingrays: Underwater world.* Minneapolis, MN: ABDO Publishing.

Cousteau, P., and D. Hopkinson. 2016. *Follow the moon home: A tale of one idea, twenty kids, and a hundred sea turtles*: San Francisco: Chronicle Books LLC.

Fleisher, P. 1997. *Webs of life: Coral reef.* New York: Benchmark Books.

George.,T. C. 2000. *Jellies: The life of jellyfish*. Brookfield, CT: The Millbrook Press.

Gibbons, G. 1995. *Sea turtles*. New York: Holiday House.

Head, H. 2013. *Amazing life cycles: Fish.* London: Octopus Publishing Group.

Landau, E. 1999. *A true book: Jellyfish*. New York: Children's Press.

Landau, E. 1999. *A true book: Sea horses*. New York: Children's Press.

Marsh, L. 2011. *Sea turtles*. Washington, DC: National Geographic Kids.

Peterson, M. C. 2014. *Coral reefs: Smithsonian little explorer.* Mankato, MN: Capstone Press.

Rake, J. S. 2009. *Beluga whales up close.* Mankato: Capstone Press.

Rhodes, M. J., and D. Hall. 2005. *Sea turtle*. New York: Scholastic.

Rose, D. L. 2000. *Into the A, B, sea.* New York: Scholastic.

Sexton, C. 2009. *Squids Oceans alive.* Minnetonka, MN: Bellwether Media.

Swinburne, S. 2005. *Turtle tide: The ways of sea turtles.* Honesdale, PA: Boyds Mills Press.

Video Resources

The Traveling Turtle: Loggerhead Sea Turtle Critter Cam. New England Aquarium. *www.youtube.com/watch?v=8nmmYAtmag0.*

Straz Center-National Geographic Live! Brian Skerry. *www.youtube.com/watch?v=FJcLR86G7YQ.*

References

Armstrong, T. 2009. *Multiple intelligences in the classroom.* 3rd ed. Alexandria, VA: ASCD.

CAST. 2018. Universal Design for Learning (UDL). *www.cast.org/our-work/about-udl.html#. WS9Nh8m1vMU.*

Darling-Hammond, L. 2010. *Performance counts: Assessment systems that support high-quality learning.* Washington, DC: Council of Chief State School Officers.

Darling-Hammond, L., K. Austin, M. Cheung, and D. Martin. 2003. *The learning classroom: Theory into practice.* Stanford, CA: Stanford University School of Education.

Edutopia. 2009. Big thinkers: Howard Gardner on multiple intelligences. *www.edutopia.org/ multiple-intelligences-howard-gardner-video.*

Edutopia. 2013. Multiple intelligences: What does the research say? *www.edutopia.org/multiple-intelligences-research.*

Edutopia. 2016. 53 Ways to check for understanding. *https://backend.edutopia.org/sites/default/ files/2018-01/edutopia-finley-53-ways-to-check-understanding-2016.pdf.*

Gardner, H. Washington Post. 2013. Howard Gardner: "Multiple intelligences" are not "learning styles." October 16.

National Center on Universal Design for Learning. 2017. About UDL. *www.udlcenter.org/aboutudl/udlguidelines_theorypractice.*

National Governors Association Center for Best Practices and Council of Chief State School Officers (NGAC and CCSSO). 2010. *Common core state standards.* Washington, DC: NGAC and CCSSO.

NGSS Lead States. 2013. *Next Generation Science Standards: For states, by states.* Washington, DC: National Academies Press. *www.nextgenscience.org/next-generation-science-standards.*

Nelson, L. L. 2014. *Design and deliver: Planning and teaching using Universal Design for Learning.* Baltimore, MD: Paul H. Brookes Publishing.

Salend, S., and C. Whittaker. 2017. UDL: A blueprint for learning success. *Educational Leadership* 74 (4): 59–63.

Tomlinson, C. A. 2014. *The differentiated classroom: Responding to the needs of all learners.* 2nd ed. Alexandria, VA: ASCD.

Wiggins, G. 2006. Edutopia: Healthier testing made easy: The idea of authentic assessment. *www.edutopia.org/authentic-assessment-grant-wiggins.*

Wiggins, G., and J. McTighe. 1998. *Understanding by design*. Alexandria, VA: ASCD.

Index

Page numbers printed in **boldface** type refer to figures or tables.

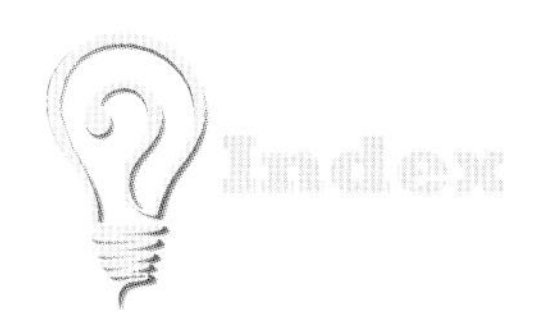

Index